PENNSYLVANIA MINING FAMILIES

Pennsylvania Mining Families

THE SEARCH FOR DIGNITY IN THE COALFIELDS

BARRY P. MICHRINA

THE UNIVERSITY PRESS OF KENTUCKY

Copyright © 1993 by The University Press of Kentucky

Scholarly publisher for the Commonwealth,
serving Bellarmine College, Berea College, Centre
College of Kentucky, Eastern Kentucky University,
The Filson Club, Georgetown College, Kentucky
Historical Society, Kentucky State University,
Morehead State University, Murray State University,
Northern Kentucky University, Transylvania University,
University of Kentucky, University of Louisville,
and Western Kentucky University.

Editorial and Sales Offices: Lexington, Kentucky 40508-4008

Library of Congress Cataloging-in-Publication Data

Michrina, Barry P. (Barry Paul, 1947–
  Pennsylvania mining families: the search for dignity in the
coalfields / Barry P. Michrina.
      p.   cm.
  Includes bibliographical references and index.
  ISBN 0-8131-1850-6 (alk. paper)
  1. Coal miners—Pennsylvania—Cambria County—History.
2. Strikes and lockouts—Coal mining—Pennsylvania—Cambria
County—History. I. Title.
HD8039.M62U64444   1993
331.7′622334′0974877—dc20   93-19826

For CZEDO and for LEOPOLD ROGERS,
both former coal miners,
one my paternal grandfather,
the other my fieldwork "grandfather"

Contents

Acknowledgments

I would like to publicly recognize and thank all of the people and institutions that contributed to the successful completion of this project.

Catherine Lutz provided knowledgeable advice, moral support, and openness to discussing our differences, and her helpful reviewing enhanced this manuscript and increased my understanding of anthropology.

The people of Cambria, Indiana, Clearfield, and Somerset counties opened their lives—past and present—to my scrutiny. They wanted to have their stories told: I hope that this document fulfills their needs. Other residents acted as facilitators in arranging interviews.

The library and archival facilities of the Indiana University of Pennsylvania, the Cambria County Library system, St. Francis College, and the Cambria County Historical Museum provided invaluable background information and quiet places for scholarly pursuit. I especially thank Eileen Cooper (I.U.P. curator of special collections).

The following people provided consultation in their specialties: Irwin Marcus (labor history), Waud Kracke (transferences), Ron Gresh and John Kuzar (mine safety). Neville Dyson-Hudson, Melvyn Dubofsky, and Jane Collins also provided many helpful suggestions.

Ron Gresh, Bill Piekielek, and Irwin Marcus provided careful review of chapter drafts. The office of District 2 UMWA provided me with access to local meetings, and the Pennsylvania Mining Corporation extended to me the opportunity to visit their underground facility (Rushton Mine). Elthea Florman kindly provided documents, pictures, and memories of her grandfather, Rembrandt Peale.

Amy Guenther worked closely with me in typing this manuscript and in bringing it to press. Her editing and her patience were important to me.

I would not have been motivated to take the first step in the journey toward this study without the enthusiastic and knowledgeable teaching of Sid Waldron at S.U.N.Y.—Cortland. He took time to be a mentor to me despite his hectic schedule. My success in fulfilling this project is but a reflection of his success as a teacher.

PENNSYLVANIA
MINING FAMILIES

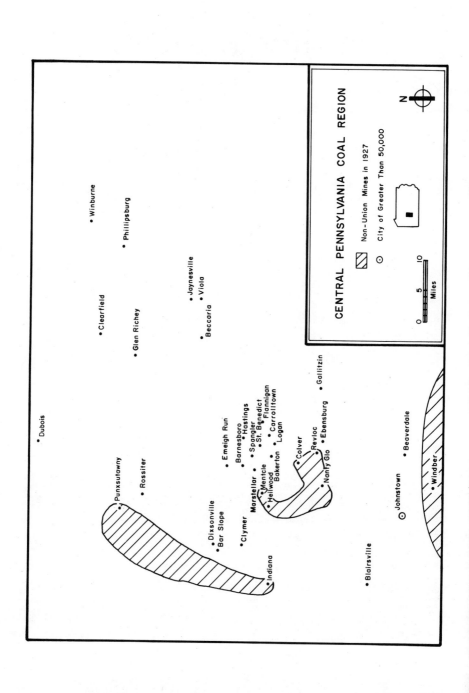

CENTRAL PENNSYLVANIA COAL REGION

Non-Union Mines in 1927

City of Greater Than 50,000

Miles
0   5   10

1

Being There, a Reflection

Other people might have started this and quit because it was too much work . . .

—*wife of a retired miner*

Because this book is likely to be read by people with varying expectations, I feel the need to explain its structure and style. I have not attempted to write a traditional oral history nor a traditional ethnography, though it contains elements of each.

By placing myself in the text as a situated thinker, actor, and interactor, I have attempted to show the nature of the investigative process. Collecting and analyzing the data was not an encyclopedic project in which I simply wrote down and considered facts from sources of indelible, objective data. Rather, I interacted as a unique individual, filled with expectations, beset with frustrations, confusion, and dilemmas, and troubled by questions involving epistemology and ethics. I feel that how I thought, acted, and reacted are important information for the reader.

Several places in the text (most notably chapters 1 and 10) have what one reviewer called a "flow of consciousness" flavor to them. This is a part of my technique for situating myself "in the field." Likewise, I have included reflexive material throughout the text. Some readers may prefer a single chapter for this purpose, but I chose to follow one longstanding style in ethnography, which includes the work of Gearing (1970), Dwyer (1982), Dumont (1978), and Jackson (1989). I feel that these reflexions concern themselves almost entirely with issues of epistemology or ethics. An example of this style concerns itself with the issue of truth in oral histories; I have addressed this in several places in the text (chapters 1, 3, and 10).

The author receives a lesson on the pick in front of the former portal of Sterling No. 2 Mine. History is more than immutable facts.

My method included negotiating with my informants in an attempt to reach an understanding. Agar (1982) has referred to this as matching horizons with the natives. When I felt I understood an issue of the natives, I then talked with them in terms of that understanding. Sometimes they would correct me; other times I would become confused by their subsequent statements. I attempted to reach a coherent understanding through this feedback. Dwyer (1982) has also spoken of negotiation between investigator and informant— each coming into the interaction with his or her own agenda. Chapters 5 and 6 most clearly illustrate my use of this technique.

It is clear to me that all data that deal with people are situational in both collection and interpretation. I have come to several overall impressions from my fieldwork experience, and I have noted the tenuousness of promoting them as "truth." Some readers may feel uncomfortable with this lack of certainty, but I would suggest to them that such certainty is illusory, that there are at best only facets or renditions of the truth.

As I stepped into the church basement I could smell the sweetness of freshly baked bread and melted butter. I faced a scene of bustling activity; gray-haired women moved rapidly with bags in hand toward parallel rows of long tables. Twenty or so of the fifteen-foot-long tables held piles of bread loaves and rolls protected by plastic bags, and waxed milk cartons now filled with pirohy that dripped with butter.

It was the Lenten season, and the orthodox rite Catholic church's women's sodality was preparing traditional Lenten food: paska (braided, sweet bread), nut rolls, poppyseed rolls, pagash (cheese and potato-filled bread), and five types of pirohies (dumplings): cheese and potatoes, cabbage, sauerkraut, cottage cheese, and prune.

I had made five preliminary, one-week trips to this area prior to beginning fieldwork. During this time I had renewed acquaintances with many people whom I had known prior to leaving twenty years before. I made many new contacts as well. In looking for housing in November 1987 for my move in January, I discovered that rental houses were rare. However, Leah and I were able to find a small house through the help of friends. It was located in a town three miles from my former hometown. This became an ideally centralized location for interviewing in the surrounding former mining towns that were strung out along creeks and rivers in northern Cambria County.

I came to my field site with conflicting desires: on one hand I sought to conduct an academic study, and on the other, I wanted to escape academia, to live in the past both geographically and temporally, for I had endured a demanding three semesters of coursework, teaching, and preparation for my qualifying exams. So it was that I came to the region looking for the rural setting of the 1950s. In many ways it still exists there. One can spend an afternoon picking wild berries among the sounds of nature, exploring old strip mines, meeting adolescents who say a genuine "hello" to adults. Many residents still sit out on their porches calling "hello" to passersby, revere religious piety, and leave their doors unlocked. Even the acrid smell of coal smoke that billowed from many of the townspeople's chimneys during the winter suggested a bygone era.

I submerged myself in what I imagined as a pastoral setting, acquiring an old pushmower to cut my grass, hanging clothes on the backyard clothesline, foraging for berries and mushrooms, joining long-standing clubs, and working at the church festival.

At one table a woman prepared balls of cheese and mashed potatoes mixture to stuff dumplings. She recognized my last name. Her uncle had married a Michrina about fifty years ago—they now live in California. Connections among people seemed important here. She wanted to know how well I knew them—she writes to them every Christmas.

I ignored changes that had occurred in the past thirty years. For example, I overlooked the aging process that had affected the population; now the oldest generation represented the major portion of the residents. I also tried to overlook the intrusion of mass society that manifested itself in decaying and disappearing businesses in the small towns, the sprawl of housing and businesses outside of town, the large numbers of all-terrain vehicles, the recent advent of cable T.V., and the fast food marts. Also missing from the area were structures from the past: the Patton Clay Works, the Carrolltown Convent, the mine office building in St. Benedict, the No. 9 mine refuse pile, and the Barnesboro and Portage movie theaters. I also tried to ignore some of the defects from the 1950s such as "sulfur creeks" that ran orange from the iron dissolved in the acid drainage of mines, and new problems such as the growing ranks of welfare recipients.

How frequently do we ethnographers describe what we *need* to find in our field situations: peace, simplicity, friendliness, altruism, and solace? I know from reading Clifford (1986) that others have succumbed to the need to salvage the pastoral culture, though their motives remain mostly hidden. When this has occurred, the people, discourse, and practices have been placed in what Clifford has called a "present becoming past" (Clifford 1986:115). He suggested that we resist the impulses of the ethnographic pastoral by opening ourselves to different histories. In my case, I tried to remain open to people's voices—the voices of mine labor that told me of the history of the labor struggle, hardships, emotions, processes, and perceptions—a realistic history underlain with both obvious and hidden aspects of power. Without the insights I obtained from reading Clifford, would I have gained this perspective?

I walked to the counter that looked in on the kitchen. Ten or so women were busy cooking. I requested my order, and a sweet-natured old woman guided me to a pile marked with my name. My request fit into two butter-stained grocery bags. I told her that I was so glad that she and her friends had gone to all this trouble to prepare such rare treats. Her eyes smiled as she told me that she was glad to make so many people happy.

My study had begun to emphasize the memories of the oldest generation of coal miners as a consequence of several factors that emerged during the fieldwork. I had begun by talking to the oldest generation of miners and their wives in an effort to help lay a historical foundation for current behaviors and attitudes. I had hypothesized that years of living under oppressive economic conditions such as underemployment, physical danger, and militant repression of rights would shape working class psychology both directly and through intergenerational relations such as childrearing. I sought intergenerational continuity. Instead, I found great intergenerational variability due to the movement of many miners' sons out of the industry. The boom of the 1970s had attracted sons of farmers, merchants, and teachers while many miners' sons had been encouraged to move out of the area to obtain an education or a skill.

I had also noted that modern changes such as mass communication and mechanized transportation had apparently affected the more recent generations of residents who had grown up with the television and the automobile. They were less isolated and less unique than their predecessors.

These older residents impressed me with the unique stories they had to tell and with the emotional fervor by which they told them. Although I had grown up in this region and had had mining grandfathers, I had not learned these stories. The older residents and I colluded in making the story known—they with their eagerness to tell and I with my need to learn.

But what of participant-observation? What reason could there be to limit my description of the behavior I had noted once I had been accepted as a native? A colleague has suggested the possibility of psychological, political, or ethical reasons for eliminating this practice or the data thus obtained. Although I felt some anxiety in playing the part of an accepted member of the region, it was not a state that prevented my participation. Nor did I see devastating political consequences in reporting or withholding descriptions of natives' natural behavior. For me, it was an ethical decision to exclude participant-observation. I was most often perceived as a neighborhood son and was invited into social circles with complete trust. These people were trusting, some were lonely and in need of therapeutic listening and reflection. Sometimes people told me accounts that resembled those I had heard in environments in which I had been a counselor. I am certain that they would have rejected my proposal to record and report the interaction between them and me or that among themselves. To collect and report such data appeared to be deceitful; I discontinued the practice.

With regard to any oral history epistemological questions arise. Is there a continuity over time in the perception of former events? What may give shape to memory? How does the process of taking oral history affect memories? I will address these in more detail after the reader is familiar with the content of the history. In recording some events, such as the incidents that occurred during the 1927 strike, I sought accuracy and compared informant accounts with the limited archival accounts that were available. In other instances I was more interested in analyzing the discourse for values, opinions, feelings, and practices. Here, I assumed a constancy over the person's lifetime.

What follows is an analysis of expressed memories of the oldest generation of central Pennsylvania's coal miners and their wives. The topics include the great strike of 1927, the time without union, work practices and values, and mine danger. Why this selection of topics? I like to think that the topics were chosen by the informants themselves; I found these topics to be of major interest among the first thirty or so people whom I interviewed. But as Dwyer (1982) has suggested, I may have influenced the selection of these by my enthusiasm to certain accounts or by my lines of questioning.

My approach to analyzing the change in work practices was heavily influenced by my reading of two texts while in the field: Goodrich (1925) and Dix (1989). These works emphasized the militant aspect of Eastern coal miners. Although I maintained my proposed emphasis on the analysis of emotions, I changed from a psychodynamic approach involving the notion of defenses against intrapsychic pain to the study of emotions as practices. I made this intellectual turn largely as a consequence of reading two volumes published during my fieldwork and my writing: Lutz (1988) and Abu-Lughod and Lutz (1990). The former questioned the ethnocentric bias associated with the psychodynamic model of emotions while the latter emphasized the politics and power relations associated with the emotional practices of a culture.

Although this study strongly resembles an oral history, it also contains numerous references to the present that pertain to the collection of memories. I write of attempts to impose my own agenda with regard to work practices, I describe the reaction of informants to chapter 5, women's practices with regard to talking to me of danger, current assaults on the residents' dignity. I tell of my efforts to seek an ethnographic pastoral, describe how miners perceived me, and write of culture-constituted emotional transferences that I felt with natives. I have dispersed these analyses throughout the text, demonstrating that the present is related to the collection of memories of the past and their analyses.

What I will suggest in several chapters of the text is that emotional practices, emotion-related perceptions, and emotion-laden morals that either arise in a working class or are reproduced in that class are internally acceptable, preferred, even demanded by the members of that class to fulfill a need for cultural meaning. Yet their existence and persistence can have detrimental consequences for the members of that class. These practices, directed within the group, may take attention away from the owning class's exertion of power and may add moral fervor to the blaming of "insiders"— i.e., strikebreaking miners, fellow miners, and miners' wives— for conditions that might logically be assigned to the coal operators.

This thesis aligns itself with that of Bourdieu (1984) in which he suggested that the members of a class internalize a classificatory system that continuously transforms necessities into strategies, constraints into preferences. I did not use Bourdieu's theory to formulate a hypothesis to test at the outset of my study. Rather, I read his book late in the fieldwork/analysis process and recognized the similarity in his theory and my own thesis. I also will show via subsequent analysis in chapter 9 that mining families used the self-image of the "little man" to modulate acceptable emotional practices in an effort to maintain their dignity.

Perhaps the most important conclusion of this study is the relative passivity of the mining community with regard to relations with the company. My findings are not in accord with the stereotypical picture of the rebellious miner or with descriptions presented by Dix (1977, 1989), Goodrich (1925), and Singer (1982). As total outsiders, these investigators may have projected their own feelings and values into their analyses.

The non-militaristic picture of the coal miner that emerged from my work agrees with the conceptions of Kerr, Dunlop, Harbison, and Myers (1960) and with those of Bodnar (1982, 1985). According to Kerr et al. (1960:187): "The industrializing societies universally come to contain, to control, and to redirect the responses of industrial workers to the transformation of society"; and (1960:190): "Workers have proved themselves much more adjustable to the impacts of industrialization on their technical and social skills, and much more agreeable to the imposition of the web of rules, than was once suspected." These authors see themselves as having "redefined the labor problem as the structuring of the managers and the managed under industrialization rather than as the response of unions to capitalism." (Kerr et al., 1960:12-13).

Bodnar (1982) has portrayed the working class culture in early twentieth-century Pennsylvania as focused inward toward family, social groups, and religion rather than outward against workplace injustice; as feeling powerless against the world outside the community; and as acting pragmatically in maintaining a secure lifestyle. I have concluded that the search for dignity oriented the mining community either inward (as was described by Bodnar) or outward, taking action against mine owners, according to their realistic appraisal of their power.

2

Coal, a Very Hard Subject

The influences of all the years of meagre living and struggle for mere existence among these barren hills, had left an imprint on these miners and their families, that amounted almost to despair. Their women folks become old and hollow-eyed before their time. The children were found undersized, and with supplicating eyes begging for help.
—*Hylan Committee Report on Berwind-White Company's Mines, 1922*

As I traveled west from Altoona on U.S. Route 22 I could understand why the coal lands of the central Pennsylvania region were developed so late. The one thousand-foot-high escarpment was a barrier to coal transport until suitable rail lines could be wound through the foothills to the top of the plateau. This plateau is so deeply dissected by valleys that the building of branch lines to the coal reserves represented a major enterprise as well. I became curious about the early coal history of the area and sought out information in written histories and archival sources. Not surprisingly, the earliest mines were developed along main lines of the Pennsylvania and the New York Central railroads. In South Fork, G.B. Stineman's Slope Mine was opened in 1869, and the Euclid Mine opened in 1875. In Benscreek, the Sonman Colliery No. 1 was opened in 1872; the Sonman Shaft was opened seven years later (Fulton 1890).

Prior to these early ventures the region had been a farming region, hewn from the frontier during the period from 1800 to 1850. The Scotch Irish settled much of Clearfield and Jefferson counties while fairly affluent German immigrant farmers settled in Cambria, eastern Indiana, and northern Somerset counties (Riesenman 1943). It was from these farmers that coal rights had to be obtained in large enough parcels to make mining feasible.

The various means by which coal rights could be obtained is intriguing. In some cases the land was bought from the Pennsylvania treasury by paying the owner's delinquent taxes (e.g. S.R. Peale, Centre County Deed Office, 1882–1886); sometimes a flat fee was paid for all the coal rights under the property (e.g. Blubaker Coal Co., Cambria County Deed Office, 1888–1889); and in some cases a royalty was paid to the landowner on the coal tonnage when mined (e.g. Rembrandt Peale, Indiana County Deed Office, 1903).

Advance men would buy coal rights in undeveloped areas and use their influence with the railroad companies to have lines built into those areas. The Pennsylvania Railroad and the New York Central Railroad had their own coal companies: Pennsylvania Coal and Coke Company and Clearfield Bituminous Coal Company, respectively, which would often begin operation in these new coal areas.

By the turn of the century coal was booming. In 1901 Cambria County had 130 colleries, Centre County 28, Clearfield County 127, Indiana Counky 30, Jefferson County 39, and Somerset 53 (Halberstadt Map). The production figures for all of Pennsylvania for the period from 1899 to 1916 was 2.3 billion tons of bituminous coal or 58 percent of the total mined in the U.S. during those years (Coal Age 1917). I found that the Central Pennsylvania District averaged half a billion tons per year during the period from 1916 to 1920. This represented 10.5 percent of the nation's bituminous coal production for those years (Central Pennsylvania Coal Producer's Association files, 1925).

I suspected that there were political connections between the region's coal interests and the state government, and I found information to confirm it. Governor James A. Beaver was a partner in the Blubaker Coal Company, a land holding company formed in 1887, and in the Sterling Coal Company, which was formed in 1902. His partners included General Daniel H. Hastings, then adjutant general and later governor himself, and Colonel John L. Spangler who was a member of the governor's military staff (Gable 1926:202-205). Jake Stineman was a state senator and coal operator in Cambria County in the 1890s (Brophy 1955:194). S.R. Peale, a state senator from Lock Haven, speculated in coal lands in Clearfield and Centre counties in the late 1800s and later became a coal operator (Peale 1861). His son, Rembrandt Peale, became associated with President Wilson's cabinet, serving on the fuel commission during World War I and on a three-member coal commission in 1919.

I wondered what some of my informants had meant when making references such as "D seam coal." I found that five seams of coal lay beneath much of the surface in central Pennsylvania. These were, beginning from the surface, the E (Upper Freeport), D (Lower Freeport), C (Upper Kittanning), B (Lower Kittanning) and A. Because the A seam is the deepest and is characterized by the highest concentrations of explosive methane gas it has been the least exploited. The D and B seams have provided the best coal for underground mining. These seams all contain "low seam coal," the thickness varying from two feet to six feet (Geological Survey of Pennsylvania, 1888). When the seam was only two to two and a half feet high, the miner was required to remove at least two feet of "floor" in order to fit his car and himself in.

In order to find out about economic conditions in the region near the turn of the century I consulted John Brophy's memoirs (1955). He had been a miner at that time and later became the UMWA District 2 president. He stated that by 1894 there were forty thousand miners in District 2 (central Pennsylvania) of the UMWA. Their fortunes swung with the boom and bust cycles of the coal market. There was a coal strike that year, an attempt to prevent the coal operators' association from reducing wages. The strike was lost after sixteen weeks, resulting in the complete destruction of the union in central Pennsylvania. A depression followed, which threw the miners into conditions of extreme poverty (Brophy 1955:57). They suffered conditions of mass unemployment; those who found employment were given work for only one day per week, and often the mines would close down for weeks at a time. Those who relocated to seek employment had difficulty establishing a home because they had to spend everything in the move (Brophy 1955:66-68). The depression in the region lasted through 1904.

In 1904, the newly reorganized union accepted a wage reduction of 5 to 6 percent because of adverse market conditions. In 1906 a strike was organized to restore wages to their previous level. In some towns, such as Greenwhich, striking miners were evicted and company-employed deputy sheriffs and strikebreakers were utilized (Brophy 1955:196-230).

I was shocked at the bleak economic conditions that plagued the area. In 1908 and 1909 there was a recession in the coal fields, and in 1915 there was another slow-down. However, World War I revived the coal market, and by the summer of 1916 there was unrest among the miners who saw operators making a profit and their relatives who worked in city munitions plants making better wages. In February 1918, President Wilson's fuel administrator allowed a price rise to the coal producers of up to fifty cents per ton to counter inflation. No such concession was given to labor where mine wages were frozen (Singer 1982:81).

According to Singer (1982:105) the war "boom" was short-lived. Following the signing of the armistice in November 1918 there was another recession. Many marginal mines, opened during the war, shut down entirely. Others worked only a short time. With both unemployment and the cost of living high, the miners struggled to make ends meet. At that time a bituminous coal miner earned an average of 1,583 dollars per year according to the Bureau of Labor Statistics (Singer 1982:105).

By reading Singer (1982); Blankenhorn (1924); and Marcus, Cooper, and O'Leary (1989), I came to understand how conditions developing after World War I affected the central Pennsylvania region until World War II. This includes the time of the 1927 strike and the time without a union. There were wildcat strikes throughout the coal fields in 1919, including the Coral mine of the Potter Coal and Coke Company. Delegates at the national convention gave Lewis the authorization to strike on November 1 if collective bargaining failed to bring an agreement for higher wages, shorter hours, and nationalization of the mines. Despite an injunction, the miners—about 400,000 nationwide, including 50,000 in District 2 of the UMWA—struck. After a second injunction, Lewis called off the strike, and by the end of December it was over (Marcus, Cooper, and O'Leary 1989:179-81). The union's executive board accepted President Wilson's proposal for a 14 percent raise in wages but added the provision that an investigatory commission further study the wage issue (Singer 1982:88).

On December 20, 1919, President Wilson appointed a three-man coal commission to appraise the miners' wage demands. It consisted of Henry M. Robinson, representing the public; John P. White, former president of the UMWA; and Rembrandt Peale, an independant coal operator from central Pennsylvania (*New York Times,* Dec. 21, 1919). Later, the commission recommended and received a wage increase of 27 percent in the tonnage rate and 20 percent in the day rate (Singer 1982:88).

The slumping coal market and the competition from non-union fields reduced central Pennsylvania's coal production in the years 1918 to 1921. In Cambria County it fell from 20.5 million to 12.9 million tons per year; in Clearfield and Indiana counties production dropped 60 percent (Singer 1982: 135). Operators sought to maintain profitability by introducing mechanization (i.e. the cutting machine to undercut the coal seam) and by switching to open-shop relations (Singer 1982:94-102). According to Singer (1982: 124) the Harding administration argued that the government had to protect the public interest by encouraging the non-union production of coal. The first challenge to the open-shop movement came in 1922 when the miners in District 2 supported a "strike for union" aimed at organizing the non-union operations within the district—places like Somerset County, Colver, Revloc, Heilwood, and Vintondale. Among the more than 140 companies operating non-union mines in central Pennsylvania at the time were the Consolidation Coal Company, a Rockefeller interest; Berwind White Mining Company, capitalists in New York transit; the United States Steel Corporation; and Bethlehem Steel (Blankenhorn 1924:5). Berwind White employees were reported to have gone on strike because the company had abolished payment for "deadwork" (i.e. work that did not directly lead to coal being placed in a miner's car; most often this involved cleaning up rock falls from main haulageways) and had reduced the rate for mining coal from $1.28 to $1.01 per ton (Hylan Committee 1922:18).

According to Blankenhorn (1924) the operators and the state reacted in an abusive fashion to the strike. He reported mine guard brutality; Coal and Iron Police and state constabulary prohibiting the exercise of civil liberties in the towns and using brutal or intimidating practices of evictions; tent colonies for the evicted; importation of strikebreakers; and famine among families of strikers. Injunctions were issued against union organizers who tried to enter Somerset County and against picketers, and, in Heilwood, the state guard was called in (Blankenhorn 1924).

Lewis signed an agreement in August 1922, ending the strike and abandoning the newly organized fields (Singer 1982:103). Those Somerset strikers who did not relocate to work in the union fields stayed out for two years. Blankenhorn (1924: 176-92) reported that the District 2 officers decided to "take the public for a partner in the coal business." They sent a small delegation to New York City to expose the operators' tactics and the strike itself to the public. The newspapers printed their story, and John F. Hylan, mayor of New York City, sent a committee to Somerset County to witness the situation itself. According to Blankenhorn, this strategy failed in the end because the issue had been pressed to a point where only the voice of a nationally known spokesman would have been listened to, and no one took up the cause.

In April 1924 the central Pennsylvania operators signed an agreement to renew the current wage scale for three years. Western Pennsylvania had already decided to reduce its wage scale (Barnesboro *Star,* April 3, 1924; *Coal Age* 1922:584). According to the National Coal Producers Association, 1924 was the worst period in the coal industry since 1894–1896. They attributed the causes to overdevelopment and lack of demand (Pennsylvania Department of Mines). During this time of economic hardship the Ku Klux Klan began to organize in Pennsylvania.

Jenkins (1986:123-25) claimed that there were 125,000 Klansmen in the state by the end of 1924 and as many as 250,000 two years later. They represented a very broad cross section of the native-born working class of the area. They demonstrated against Catholics, the religious denomination of most of the eastern and southern European immigrants working in the state's mills and mines. Altercations took place in central Pennsylvania.

In February 1924, two Italian miners were arrested on charges of shooting into the home of the minister of the Lilly Lutheran Church. Italian miners dominated the UMWA local and responded to the arrests by suspending six Protestant miners suspected of Klan membership (Singer 1982:218). Lilly was on the main line of the Pennsylvania Railroad, an easy spot in which to organize a massive march and cross-burning. Twenty miles away lay Altoona, a city that may have had the highest number of Klansmen per capita of any city in the United States (Jenkins 1986:124). An eyewitness to Klan activities in Lilly on April 6, 1924, told me that a train from the East as well as one from the West brought hooded members to the town. The newspapers reported only a train arriving from Pittsburgh. Three young townsmen turned a fire hose on the Klansmen following a march and cross-burning. Subsequent gunfire killed three villagers and injured at least twenty more. Twenty-six Klansmen and several townspeople were arrested (Barnesboro *Star* April 10, 1924).

Singer (1982: 216-19) also described Klan activity. According to his account, the open-shop drive in the Nanty Glo region was preceded by a period of intensive Ku Klux Klan activity. The Klan disrupted the Nanty Glo Columbus Day celebration with a cross-burning in 1924. The miners suspected that the management of the Heisley No. 3 Mine in Nanty Glo was sponsoring local Klan activities in preparation for breaking with the union. They also suspected that newly elected Local 1347 president, Samuel Chilton, was a Klan member. Despite the intense activity in 1924, the Klan support quickly subsided. By 1928 the state membership was down to 30,000 (Jenkins 1986:131).

In 1925 and 1926 the operators began repudiation of the 1924 contracts. A report in the District 2 UMWA file described a tactic used by the companies. The tactic could easily be interpreted as a lockout—companies would close down operations while having their orders filled at other mines. They let their own mines remain idle until their union employees were starved into submission. They then opened their mines at a lower wage rate and under non-union conditions. When the Heisley No. 3 Mine, a Weaver holding in Nanty Glo, attempted a contract repudiation in 1925, the union workers picketed. This marked the beginning of what would be called the Great Strike of 1927 in District 2.

In a 1926 letter John Brophy indicated seeing already a dismal situation (Singer 1982:233):

I don't know of a time since the middle '90s that conditions were as gloomy as they have been during the present, and there is no prospect of improvement in sight. Scores of mines in the district have been shut down for a year, and others are working a day or two a week . . . In addition to this, injunctions, evictions and all the paraphenalia of the strike have been employed by companies against us. Local strikes, of which we have quite a number, are deadlocked hopelessly without a decision. This year I expect will be worse than last year. The union appeared powerless to stem the open shop tide. In District 2 many miners accepted an offer from small operators of steady work at the 1917 scale.

Five or six of my informants could recall some details of the 1922 strike for union, and ten or so remembered the company repudiation of contracts, particularly Nanty Glo resi-

The long history of
coal shows on the
landscape: the town of
Ehrenfeld, 1988.

dents. For most of the mining folk, the 1927 coal strike
represented the first significant political event in their lives.
Their accounts of those times captivated my attention as no
other information could.

NOTE: The chapter title comes from a remark attributed to Rembrandt
Peale in *Cushing's Survey* 8(42), July 3, 1930: "That Agency is entirely
too soft to deal with a hard subject like coal."

3

The Great Coal Strike
of 1927

Sometimes you wonder how you ever made it—leastwise I do.
—*retired mineworker*

Of the topics discussed during my fieldwork, informants
spoke of the 1927 coal strike with the most fervor. The reasons for this varied: several wanted the event to be docu-
mented, some wanted to convince disbelieving children and
grandchildren, and some seemed fascinated by the severity of
the times—as if it were a part of their identity. I found myself
fascinated by the sinister and sometimes excited quality of
their accounts, and I found myself surprised that these events
could have occurred recently enough to be remembered.

I wanted to document these accounts of the 1927 strike
and of the subsequent time without union largely because
there is so little archival documentation of those times in central Pennsylvania. I wish to do this not only to satisfy the
wishes of the informants but to show the variability in people's behavior and the long-lasting effect on many people's
emotions.

I interviewed 109 people who had lived in the area's coal
mining towns during the 1927 strike. In what follows I have
summarized mostly informant accounts describing the events
and conditions these mining families experienced. I also used
several sources of archival data. I found that in every case the
archival information verified or supported my informant
narratives. I have saved some of the richness of detail inher-
ent in informant quotations for chapter 4 in which I analyze
the emotional nature of those times.

I have taken Deryk Holdsworth's advice to investigators of coal mining communities to describe existing variability rather than merely to seek stereotypes (1989). Beset with variable treatment by the company, varying union support, and varying options for survival, miners and their families reacted in a range of ways to the conditions of 1927 and 1928. For most, it was a period of deprivation, of repression of civil liberties, and of turmoil. This represented the last strike of warlike character in the Pennsylvania coal fields, resembling in many respects the 1922 "strike for union" (Blankenhorn 1924).

Not all mines were unionized when the strike call came in the spring of 1927. The map of union and non-union mines in the front of this volume is based on informant accounts and archival records. Mining towns that were non-union, at least since 1922, included: Vintondale, Revloc, Colver, Heilwood, Mentcle, Brownstown, and all coal towns in Somerset County. Several mines became non-union in 1925 by closing down operations and reopening six to ten months later as non-union mines. This tactic of contract repudiation was successful because the company could attract many desperate miners who were grateful for a job regardless of the union status. These mines included Rochester and Pittsburgh Coal and Iron Company in Lucern, Empire Coal Company in Clymer, Emmons Coal Company in Marion Center, and two mines in Nanty Glo: Heisley Mines and Lincoln Mines.

Informants referred to the first class of non-union towns mentioned above (i.e. longstanding non-union) as "closed towns" or "captive towns." Here martial law prevailed to keep union organizers and merchants out, to prevent union organizing among employees, to reduce absenteeism at work (by police harassment), and to keep general law and order. Since there were no striking workers in these towns in 1927, there were no evictions, and the squad of company police was small. According to records of Bethlehem Mines Corporation, there were for Heilwood, Mentcle, and Brownstown the combined force of four patrolmen and a captain. They reported police actions such as destroying union literature, refusing visitor admission, and exclusion of agitators. The small squad ruled with an iron hand, intimidating residents with threats and violence. A resident remembered:

The coal company had an iron hand and ran the company towns like a concentration camp. There was a searchlight mounted on a tower at the end of the village. It was turned on after the 9:00 curfew. A whistle blew and the Coal and Iron Police would patrol the town on horses. There was no freedom of assembly. The Coal and Iron Police would stop people getting on and off the train and interrogate them: "Do you have permission? Where are you going? Who are you seeing?" . . . They would pull men out of the houses and hit them with clubs.

Through personal accounts and newspaper articles I found that some companies, such as the Sterling Coal Company in Bakerton and the Rich Hill Coal Company in Hastings, closed down on April 1 and did not offer to reopen until December of that year. Union correspondence of the time revealed another method, exercised by the majority of union mines—to offer the mineworkers a choice between working for a reduced wage or losing their jobs to imported strikebreakers. In such cases, three sources of state-sponsored militia were utilized to protect company property, to keep law and order, to intimidate strikers, and to enforce evictions from company houses.

According to my informants and to Singer (1982) there were in Nanty Glo both striking and recently non-union mines. The town itself was strongly pro-union and had a police force large enough to counter the state-sponsored forces. Trouble had been brewing in Nanty Glo since the change in non-union status of the Heisley and Lincoln Mines in 1925.

According to Smith (1986), it was possible for Pennsylvania coal mine companies to request the aid of several types of police—all armed and on horseback: deputy sheriffs, who purchased their own badges, pistols, and maces, and were most often referred to as "pussyfoots"; deputy constables, whose presence had to be petitioned for by no fewer than twenty-five taxpayers; state coal and iron police; and state police, who were the only group of law enforcers who were state disciplined and were not paid by the mine companies.

The company rationale for having the police was to protect their property and personnel (which included strikebreakers). Other duties included transporting strikebreakers to and from the mines and removing the belongings of evicted miners from company property.

Most of the people whom I interviewed used the terms "pussyfoots" and Coal and Iron Police interchangeably and often used both terms to refer to all four classes of police. There were exceptions. Some said that constables or "constablers" patrolled in the boroughs of Barnesboro, Gallitzin and Nanty Glo. A Barnesboro resident told me: "They also had state police—called them con'sta'bulls, at least I did. They stayed at the Commercial Hotel and policed the town. They'd get pussyfoots if they ever stepped off the company property . . . the constables wouldn't let you congregate downtown. You had to stay twenty feet apart. I remember them chasing a group of us teenagers in town."

In Marstellar residents told me that pussyfoots in plain clothes later patrolled in place of the Coal and Iron Police. People told me of state police activity in Bakerton, St. Benedict, Nanty Glo, Madera Shaft, and Winburne. Residents in Mentcle and Heilwood told me that they reserved the term "pussyfoots" for company spies among the miners.

In almost every mining town I heard of abuses by the police, including harassment, assault, and brutality. A St. Benedict resident told me: "I remember this one time a miner was complaining about the meat at the company store and a pussyfoot hauled off and hit him—for complaining about meat!" A Nanty Glo resident recalled: "[A striking miner] was walkin' across the tressle by Springfield [Mine] on Sunday after church, and a pussyfoot shot him in the stomach. He carries the bullet to this day!" In Heilwood a retired miner told me: "I remember when Smitty beat up Donahue and the man with one arm—that blackjack was movin' so fast it looked like he was playin' a drum." A retired miner in Dixonville informed me of an incident at nearby Bar Slope Mine: Two striking miners "decided to drive down to Bar Slope to see what was goin' on. They didn't even stop. A guy named Little took him below the eye with a shot—it came out through the back of his neck. If his brother could have got [Little] that night he woulda killed 'im." I later found an affidavit that had been filed with the governor in the archives; it confirmed this account.

There were many reports of mounted policemen chasing taunting children. For example, a retired miner relayed a story to me from his boyhood in Curwensville: "My cousin once yelled, 'Ride 'em cowboy' to a Coal and Iron Policeman, and he chased her on his horse right up into her yard. He came on like a Gestapo after her." A miner's widow told me: "Once, when I was takin' care of the store, Mickey was sitting outside—he was retarded. He called the Coal and Iron Policeman 'Pussyfoot,' and that guy rode his horse right up those steps and came right up to the door." There were also many stories of mounted policemen patrolling areas off company property. A former St. Benedict resident remembered: "The pussyfoots would ride out almost to the vets and back. The union had barracks out there . . . The pussyfoots would ride up there to irritate the people." One informant told me of a child being killed by a policeman's horse in Bakerton, and a resident of Mentcle told me a chilling tale of policemen severely beating a teenage brother:

One time someone busted the windows out of what they called the bachelor's shanty—a boarding house. At 10:00 P.M. we heard a knock on the door. It was the Coal and Iron Police: "Where's your boy? Get him up!" They interrogated my fourteen-year-old brother: "Was you down there? Put on your pants and shoes!" We were all frightened. Two big bruisers took him out in the country and beat the hell out of him to get names. He came home all bloody.

There were reports of sexual harassment of women by the Coal and Iron Police. For example, a former St. Benedict resident told me: "[The Coal and Iron Police] would get all the girls in town and do harm to them. In that respect they were bad. You had to wonder if any of the kids born around that time belonged to their fathers."

Informants told me of miners who were mysteriously shot or disappeared in Viola, Nanty Glo, and in Mentcle. A Mentcle resident described a case of disappearance that he equated with murder: "There were a lot of killings in those days but everything was smoothed over. They'd burn the bodies so there was no evidence. Once there was a fire in the barn on the company farm, and a charred skeleton was found—the same time a miner disappeared!"

The United Mine Workers of America District 2 Office filed a report in which they described a tendancy for the state police to send in violent troops to intimidate strikers following the operator's posted notice that they would begin operating non-union. These men would disappear and a more fair set of state police would replace them.

According to most informants, the company was able in many cases to have a justice of the peace for its town, making the entire justice and law enforcement system company-run. Heilwood, Mentcle, and St. Benedict were examples of this. One informant disgustedly reported: "The magistrate, the police, the state police—they were all in the companies' hands!"

On the other hand, I found that some communities, such as Bakerton, Nanty Glo, Beccarria, and Winburne, had town constables who were pro-union. This led to violent power struggles between the two enforcement agencies. In Bakerton, the constable was critically injured by Coal and Iron Police and died a few years later: "They killed my uncle—they did! He died a coupla years later. They waited until he didn't have his [constable] badge on and then they beat 'im . . . The pussyfoots set dynamite off under my uncle's outhouse late at night. When he came out with just his long underwear on they beat him up. You might as well say they ruined that man. They broke his back and bruised him all over . . . He laid unconscious for two days in the Cambria County jail."

According to newspaper accounts, the constable for Becarria Township was beaten unmercifully while attempting to arrest two Coal and Iron Police as ordered by the justice of the peace. Later he and two of his children were assaulted by tear gas bombs by two Coal and Iron Policemen.

Residents of Marstellar, Bakerton, Winburne, and Nanty Glo told me that the police were required to transport accused persons more than ten miles to bring them before a justice who was sympathetic to the company. In some cases, the defendant was left to find his own way home. This must have been a common practice in Winburne because there was an affidavit filed with the governor that described the practice; also, a Winburne resident told me: "That son-of-a-bitch Blackie got me when I was on the picket line. There was some men goin' to work, and he accused me of startin' a fight with them. He took me twelve miles to Oceola Mills to the J.P. and then told me I had to walk home."

People told me that in St. Benedict, Marstellar, Heilwood, and Gallitzin guard stations were built in locations throughout the town. I received many reports of the police patrolling on horseback. Older residents informed me that in Heilwood and in Nanty Glo, Bethlehem and Heisley companies, respectively, erected a tower with a spotlight that illuminated the whole town.

As might be expected, the activities of children varied. Many reported that they were kept home by parents, away from strikebreakers, Coal and Iron Police, and the possibility of violent confrontation. Others described their freedom to be in parades, rallies, picket lines, police guard houses, union meetings, and downtown. Some described the action as a source of excitement.

The lives of many mining families were radically disrupted as they found themselves without income or housing. Striking miners in most instances were given a two week's notice to move out or to be evicted. A Nanty Glo miner's wife told me of what the eviction was like: "Durin' those days they didn't care about throwin' you out. I had a red-hot fire in my cookstove, and [the pussyfoots] still picked it up and took it out. They'd give you a notice that you had to get out or be throwed out. I knew they was comin'—that's why I started that fire in the stove . . . They put the furniture out in the yard, and you had to move it yourself—the union moved you." There was considerable variablility in survival options available to these unfortunate people. The following descriptions are based on the accounts of these informants.

In a few cases miners were allowed to live in company houses if they could meet the rent payments. This was true at Sterling Mines in Bakerton, at Logan, at Winburne, and Glen Richey. In the cases of Sterling, Glen Richey, and Logan, the mines shut down in April, and in Winburne company-built barracks housed the strikebreakers for Peale, Peacock, and Kerr. This company also built strikebreaker barracks in Spangler and Emeigh Run, but miners were evicted in those cases. Likewise, about twenty strikebreaking miners without family were housed together in one large house in St. Benedict. This may have been done in hopes of an early settlement with the union but, in any case, was probably aimed at preventing deterioration of company housing.

Informants agreed that all mines that were non-union since the 1922 strike allowed their miners to remain in company houses as long as men worked and paid their company-store bills. A former miner described conditions in non-union Colver: "In scab days . . . if you didn't show up for work the pussyfoot would come to your house to check on you. You had to be in bed and half-dead, or they'd drag you to work."

In most areas there were privately owned houses that adjoined the mining town or were owned by farmers. These often already housed miners even prior to the strike. Barnes and Tucker was reported to have sold its company houses in North Barnesboro to individual owners prior to 1927. Some homeowners took displaced mining families into their houses, at least temporarily, some adding rooms onto their houses for this purpose. Some homeowners left the area and rented out their houses to the homeless.

According to informants the union also provided make-shift housing in the form of barracks in Marstellar, St. Benedict, Bakerton, Tipperary, Gallitzin, Clymer, and Nanty Glo. These rough-hewn wood shanties with no inner walls were long enough to house four families in one or two rooms each. A Marstellar resident described to me the barracks built there: "The local union built a beech-plank barracks for people. Four or five families lived on one floor. There were no baths or running water. There were cracks this big [fingers indicating one-half inch] in the walls and in the floor. They took pasteboard boxes and put them up against the walls. It took years to get them outta there."

Tents were also used in Nanty Glo and Gallitzin. A Nanty Glo widow told me: "[My father] bought a great big tent and put it up in the back yard. Four families lived in that one tent."

Another option exercised by some families was to live with relatives. The move might be to a nearby town or a far-off city. Near St. Benedict, land was loaned to families for building makeshift shanties. Regardless of the living alternative, most evicted families required storage space for some of their belongings. In St. Benedict, Nanty Glo and probably elsewhere, belongings were stored by the union on the second floor of the miner's hall.

Striking families were able to get coal to heat their homes by picking it from the boney piles or by trading labor at a local house coal mine for a wagonload of coal. In Bakerton, they were able to get it from a union-run country bank mine.

The union also paid members for awhile in most areas. Bakerton, and probably other locals, were reported to have too little money. In St. Benedict, people reported that one or more of the union officers absconded with union funds, thus denying them a source of money. One informant claimed that the lack of monetary support from the St. Benedict local caused his family to move during the strike from Spangler to Revloc where his father worked in the non-union mine. One man reported that his father was denied money despite his union loyalty because of a loss of records concerning his transfer of locals.

Surplus flour was also distributed by the union, and the Loyal Order of the Moose provided charitable help to the towns of Gallitzin, Cresson, Portage, South Fork, Nanty Glo, Beaverdale, Barnesboro, Bakerton, Patton, and Hastings.

Local residents provided cooked food in St. Benedict, and farmers provided produce via several channels: gleaning, exchange for labor, gifts, and stealing. A Cambria County farmer told me: "We'd be sackin' potatoes, and as we were harvestin' from one end of the field people would be stealin' potatoes from the other end." In towns such as Marstellar, Nanty Glo, St. Benedict, and Barnesboro privately owned stores would extend credit to striking miners, a practice leading to many unpaid debts. The daughter of a miner/grocer told me: "My father mortgaged his home to keep [the strikers and their families]. Some never paid 'im back. People owed him two thousand to three thousand dollars."

An option for those with their own house was a garden. This was the single most talked-about survival technique. It was exercised even among families in which the miner worked because his mine was non-union in 1927, and by strikers who eventually returned to work during the period without union. These miners were working only one or two days a week, and survival was still tenuous. Some among them raised rabbits, pigs, and chickens for meat. A cow was certain to be kept for milk by those renting farms, and even by some people living in town, where milk was shared with neighbors.

Hunting and gathering provided rabbits, deer, groundhog, five types of berries, and several varieties of mushrooms. Meat, vegetables, and berries were canned for winter, and mushrooms were dried on strings.

Only families of strikers had to deal with the loss of the breadwinner's salary. In towns that had been non-union since 1922 the men worked short weeks of one to three days. The same was true for strikebreakers at striking mines. There were at least three classes of strikebreakers: the "company man" who did not go out during the strike, "imported scabs" who were brought from other counties or states, and labor enticed from the local farming community. The imported men were often men who had been displaced by mine closings or who had remained loyal to the union during the 1925 breech of union contract by companies like Heisley, and Rochester and Pittsburgh, but who had reached the limits of their tolerance for destitution. Some companies tricked union men into agreeing to move and work in their mines by running advertisements promising good working conditions and no labor problems. A St. Benedict miner remembered hearing of Peale, Peacock, and Kerr Company advertisements in Pittsburgh newspapers: "Peale would advertise: 'four feet of clean coal' to attract scabs. Number 10 [mine] used to be between three foot six inches and three foot eight inches." Another former miner told of an episode in which he and his father were tricked into scabbing by a false advertisement in a Johnstown newspaper:

Mind you, the union was already broken in Pittsburgh in 1926. Some companies would place ads in the papers . . . "No labor problems, houses available, will furnish transportation." In late March 1926, we got a couple of cars to go to Johnstown . . . When we got to Johnstown they piled us into the back of Dodge trucks with wire cage backs and took us to . . . Colverdale. The company had erected about one mile of fencing around their property, and they had barracks built—that's where they put us. That night people from outside yelled "Scab! Scab!" all night. My father said to me: "See, Sonny, what we got ourselves into? We're good union people." We had to go to the mines the next day . . . The second night in those flimsy barracks, the union people started shootin'. I heard a "bang" then "tinkle, tinkle, tinkle". A few BB's had found our barracks window—just one and one-half feet from my head.

They give you dog tags, too. If you wanted, you could sign up there. They would even move your family and furniture if you wanted. There were some miners from West Virginia, Tennessee, and Indiana—all at that one mine.

The next night a man who had spoken earlier to my father helped me get by the guard. Once we got out we had to run like hell. [When you got out] the local union would give you food and one dollar to help you out.

In Nanty Glo, with five major mines and a breach of contract from two of them in 1925, there was a situation of mostly local strikebreakers.

In most cases the families of striking miners had sources of income to pay for clothes, rent, and supplemental foods. In some cases children worked, sometimes sending money home from a distant city. Some men worked for local farmers. Other men would leave their families for a time to seek factory work in cities. Some of these men, taxed to the limits of endurance, abandoned their families. A former miner lamented to me: "My father left when I was fifteen. He left six brothers and sisters for me to raise. He studied violin and piano in a Bicksburg Germany monastery [when he was young]. He wasn't much of a miner. He didn't want to live in the coal mines, but this was my mother's home. He was educated enough that he wouldn't put up with it." As time wore on, men were tempted to go back to work—to "scab" for awhile. An informant offered to lend me a ledger book he had found that recorded the company house rental and company-store purchases at the Webster Mine in Nanty Glo. This lucky happenstance gave me an idea of the transience of scab miners in 1927 and 1928. The figures indicate that between April 1, 1927, and October 1, 1928, there were 110 temporary rentals, the largest proportion of transients (57 per cent) staying six weeks or less. These were most likely families. Lone men looking for work would likely to be even more transient. A former truck driver attested to this: "I hauled more of those strikebreakers back to Pittsburgh! They were strung out along Route 22, and I took a load back every time I made a trip. All they wanted was to work long enough to get some food and clothing."

Another means of scabbing was to take a mining job in a different town. One man told me: "I'd see people from [St.] Benedict and Clymer at Nanty Glo and people from Nanty

Glo at [St.] Benedict. They told me, 'You don't work in your own home town during a strike.' " The men at Sterling Mines in Bakerton and Rich Hill Mine in Hasting were given the option of working at their reopened mine with a lower wage contract. At Rich Hill, the majority of men voted in December 1927 to go back to work. At Sterling men drifted back. A resident of the area described to me the reaction of some men who felt desperate enough to go back to work after months of striking, pointing out that people were starving and needed to eat.

By late September 1928, the Central Pennsylvania Bituminous Coal Operators announced an open shop policy. Most men considered this a sign that the union was broken and exercised the option to return to work as a non-union worker. Many to this day refer to this as "scabbin' " and as "scab times."

For their psychological well-being, men found outlets for forgetting the deplorable conditions they suffered. Strikers in Barnesboro gathered at the Moose after picketing. Union men played basketball at the union hall in St. Benedict and in Nanty Glo. The Penn-Industrial Baseball League dissolved, but young union men formed their own league. This was not only recreation for the players—it provided entertainment for large crowds of people of all ages and gender.

It is safe to say that any mines that were operating in 1927 were picketed. Court injunctions brought by companies stipulated that any picketing had to be carried out away from company property, and that pickets separate such that each group consist of no more than two and that such groups be separated by thirty feet. These regulations were enforced by state police, special deputies, and Coal and Iron Police—all generally mounted on horseback. Although some report being denied the right to yell "scab," most reported that picketers yelled that and more vile names at the strikebreakers attempting to go to and from the mine. The picket line was the site of most confrontations with the police. Arrests were reported as well as cases of police riding horses into the pickets, swinging clubs. For example, a Nanty Glo man told me: "We were on the picket line and the pussyfoots come in on horses swingin' sticks. They hit my father over the head." A Nanty Glo woman told me: "Pussyfoots beat picketers. They split one man's head open and beat a pregnant woman."

From April 1 until early June three companies (Clearfield Bituminous Corporation; Peale, Peacock, and Kerr Mining Company; and Pennsylvania Coal and Coke) tried to negotiate new contracts with the union. According to informants these companies requested that miners continue to work in their mines at the old contract wage (called the Jacksonville scale) during negotiations, but most of the miners continued to strike and to picket.

The union provided transportation to different mines for picketing. People reported going from Marstellar to Nanty Glo and to Bakerton; from Barnesboro to Nanty Glo; from Clymer to Barnesboro; and from St. Benedict to Bakerton. Bakerton and Nanty Glo were sites of most intense confrontation with operators. A man from Barnesboro told me: "Nanty Glo was a strong union town. We went to picket there in 1927 at their big mine there—Heisley Mine. One guy got juiced up—ya know—got loud and a pussyfoot roughed him up. There were maybe seventy to eighty men who left from Barnesboro—the union was havin' a special celebration." Bakerton natives told me of a solid picket line of a thousand people stretching for a mile from the center of town to the Barnes Mine. Some who traveled to other towns to picket would include a visit with relatives.

Union parades and rallies were reported in Barnesboro, Gallitzin, Nanty Glo, and Bakerton. The Heisley Coal Company's plaintiff's bill for an injunction called a Nanty Glo parade "a show of force."

Informants reported seeing no women on picket lines in St. Benedict, Emeigh Run, Spangler, or Marstellar. They were reported to have picketed at Bakerton, Nanty Glo, Winburne, Beaverdale, and Tipperary. In fact, some informants stated

Nanty Glo's Heisley Mine had two-family houses. Picketing here in 1925 set the stage for the 1927 strike. Photo from top of a boney pile, 1989.

flatly that women would not have done that in 1927. I have been puzzled by this discrepancy but have noticed that women on the picket line seemed to be observed in towns of most severe conflict. Conflict was relatively mild in St. Benedict where the paternalistic Rembrandt Peale tried to reason with striking miners. Since Peale maintained a residence in St. Benedict, and since Emeigh Run and No. 9 in Spangler were nearby Peale coal towns, conflict could possibly have been mitigated there as well. In Bakerton, the feelings grew hot between Richard Todhunter, the reputed hard-nosed general superintendent of Barnes and Tucker, and the strikers. Nanty Glo strikers were embroiled in a conflict with Heisley Coal Company that had stretched on from 1925. Winburne had several mines of Peale, Peacock, and Kerr Company and of Pennsylvania Coal and Coke. In this case, the town was far from Peale's headquarters in St. Benedict and probably not visited by him. It is possible that under less volatile conditions the cultural norm was for women to avoid the picket lines, while under more inflamed conditions norms were broken.

Women were known to make catcalls at strikebreakers and police from their houses in Spangler and Barnesboro. A Barnesboro native said that "Women would open the house door and yell at the pussyfoots 'Meeeowww', or they'd hear a scab comin' home from work, open the door and yell, 'Meeeowww!.' Some of those women were rugged and the kids too." If necessary, women were capable of taking physical action against the police. A hot poker and boiling water were reported weapons. A resident of Cardiff said to me: "My mother threw a pot of hot water on a pussyfoot's horse. He said, 'You have to get out of the house!' and started up on the porch."

The daily work of women prior to the strike was arduous enough—large families to care for and the concern with ubiquitous coal dust. They also made their own bread and noodles and did their own canning. When family income was reduced, women could also be found gleaning coal from coal piles to carry in sacks on their backs. A former truck driver mentioned seeing this activity: "Down at the No. 24 Mine near Dixonville I'd see those Slovak women filling up burlap sacks with coal, and they'd sling them over their shoulder and walk up that hill without stoppin'." Many families took in boarders, and it was the woman's job to clean and cook for them. Older female children were likely to work either locally as cleaning help or clerks or in distant cities. One woman spoke of the many moves made in an effort to find work without "scabbin'." She described eight moves in 1927 alone, and twelve moves in the five year period from 1927 to 1932. She said:

I got so tired of cleanin' up other people's places—there were some pretty messy places, too!—people who would just move out and not clean . . . I did all the packin'. And we rubbed clothes on washboards, did our own bakin'. Everything you could get ahold of you canned. I'd buy groceries ahead and stored it all up when there was a strike comin' on. I had it stored in the bedroom. I had two children during those times—one born in 1927 and one in 1929.

As I wrote this chapter I was reminded of the suggestion by White and Linstrom (1989: 3-42) that the study of war stories (e.g. stories by Pacific Islanders concerning World War II) may provide fruitful data for cultural analysis. Memories may provide a connection between past and present—a continuity of perspective. During my study of memories I served as an intent listener and a devoted scribe to those who may have feared that their stories were not appreciated or would be forgotten. It also has provided us, the anthropologists, with forms of collective representation that conceptualize issues such as power and vulnerability.

I have found that, despite the diversity of conditions and of survival options, certain patterns of commonality became clear as I read these descriptions: the opposing categories of scab and union man; the universal issues of deprivation, turmoil, conflict, and powerlessness; the long-lingering emotional issues. These will emerge more distinctly in my analysis in subsequent chapters.

I hope that this synopsis of events and conditions surrounding the great coal strike of 1927, as remembered by its participants, will serve as a basis by which readers appreciate issues in following chapters: the work ethic of the handloading generation, the long-lasting emotional consequences of those times, the conflict of interests and the loyalties of workers to some mine owners.

4

Emotions Related to the Great Strike

Whole families had reason to be upset and discombobulated.
—*former miner, remembering the 1927 strike*

As I interviewed people concerning the strike and "scab times" I was aware of emotions either being expressed or described. I sensed some regularities, but it was not until I began writing that I understood the significance of this information. During the interviews I had only occasionally asked how people felt about certain situations. Rather, the data emerged as people spoke in their own words about the events and conditions of those times.

How are these emotional practices from the last great strike and subsequent time without union to be understood? I will argue that some of these practices focused attention and moral indignation on agents of the company, rather than on the owner. As a consequence the power imbalance between workers and owners was strengthened.

Among the most intense feelings of the survivors of union families, expressed directly to me or in the tone of their language, were contempt and resentment for strikebreakers. By contempt I am referring to disdain, to looking at another as worthless, negligible, or distasteful. It was directed against the strikebreakers for several reasons: for stealing their jobs; for hurting their pride in workmanship; for breaking the union.

I could recognize it in the language used. Besides the term "scab," the informants would also refer to them in animal terms, as if to indicate that strikebreakers were less than human:

It was like haulin' animals in those trucks—way the Hell from Chickoree. They were breakin' the union, workin' in there.

They moved people in here at night in trucks. People got paid by the truckload. They would bring a bale of straw in with the scabs, and they slept on it in the company houses.

They brought Hungarians in from West Virginia. We called 'em West Virginia Snakes.

They brought scabs in dog catcher wagons. Some of them never seen a mine, I don't think.

The fact that untrained men had come in and mined coal seemed to have injured the pride of union families. I deduced this from deprecating remarks made about the competence or endurance of the strikebreakers:

They brought Mexicans in to work. After one day you'd see them walkin' down the road. They didn't like it.

When the [25 Mexican bachelors] left [St. Benedict], they left behind their mining buckets. They weren't goin' to have anything to do with the mining business again.

I was especially intrigued by two jokes told about strikebreakers. This involved ridiculing of strikebreaker intelligence, and seemed to be related to the miners' own pride in their work:

There was a funny story about [Mexican strike-breakers]: A boss found some Mexicans leavin' the mine. He asked them why they were leavin'. They said, "We have to find what kind of gun you use to shoot down the coal."

They got young boys from Ebensburg [to use as strikebreakers], too—boys who never saw the inside of a mine. There's a story that two of these boys came to their place in the mine and saw the cut along the bottom of the coal and all the bug dust [coal dust from the cutting machine]. They worked for awhile and then looked for someone to "cut us some more coal." They just loaded the bug dust—they didn't know you had to shoot the coal. They thought that you just kept cuttin' it like wood.

The men who relayed these stories to me did so because they wanted me to have some funny anecdotes for my book. It

seemed to me that the narratives lacked the emotional quality they likely held in 1927. When I heard the first of these
stories I was struck by its similarity to a story heard and relayed by Goodrich (1925:110). I believe that Goodrich erred
in treating it as a factual story: The foreman "had given the
[foreign] man his place and had hung from the room the two
sights that were to give the line of direction for the room.
Next day another miner happened to come by and saw that
there was only one sight hanging up: 'What's the matter,
John? Where's your other sight?' 'Me keep it safe,' explained
the greenhorn. 'Use it when that one gets broke.' "I wonder if
this story was originally fabricated out of contempt for foreign strikebreakers.

In addition to describing strikebreakers as incompetents
and as animals, union loyalists from that time also described
them as thieves. This is recorded in the first line of a song
written by two women from Nanty Glo in 1926 (Singer 1982):
"Oh Stranger, why did you come here to take our homes and
bread away?" Informants used similar sentiments. For example, a Marstellar woman told me: "It felt like they were taking
the bread out of your mouth at the time." Some union men
felt robbed by strikebreakers in other ways. For example,
later, when the strikers considered the union to be broken
and went back to work, they found that the strikebreakers
had been given the best jobs. A retired miner from Gallitzin
told me: "Of course, the scabs got the choice jobs." A retired
Marstellar mineworker said: "The sons-of-bitches got the best
jobs— like motorman. The strikers got the worst jobs." Several men mentioned an even later loss: "What was bad—
when we realized that the strike was broken, when you went
back [to work] you was a 'S.O.B. union man.' The scabs got all
the good jobs [at Pennsylvania Coal and Coke]. Then when
seniority come in, who d'you think got seniority?—the scabs!
Union men didn't get credit for seniority. They started you
out as a new man."

A retired mineworker brought me a copy of the company
seniority list for his mine. He pointed to the top of the list and

said to me: "There's the guys who broke the union." Job seniority within the union became important for job-bidding, and the former strikers found themselves robbed of favorable position. By an ironic twist of fate, strikebreakers or their sons also became union officers when the unions reorganized in the 1930s, a condition which some loyal union people found galling. For example, one woman told me he "broke the union, and yet he joined the union when they reorganized. It didn't go down good, but still they was in there. People still called them 'scab'; there was hard feelin's." A man from a different town said: "When the union started back up, the president [of the local] was a scab. [He] had a big mouth!" Informants not only described strikebreakers as thieves with regard to the striking workers but also with regard to the company: "There were two Mexican families come in. They lived up here near us. They had Overland touring cars. The company give them credit out of the store. They run up a bill of two thousand to twenty-four hundred dollars. Then one day they just loaded up their Overlands and headed towards St. Joe's [Church]. Nobody heard from them after that." Another said: "There were some men who worked for a little while. They got all they could from the company store and then quit." I was surprised when glancing at a copy of the March 16, 1927, issue of the *United Mineworker's Journal* to find that the notion of strikebreakers as incompetent workers was bolstered, if not created, by the UMWA leadership. On page five of that issue was a quote of John L. Lewis during a visit to non-union coal fields near Pittsburgh: "A large percentage of these strikebreakers are negroes imported from southern states who are shiftless and inefficient workmen. The remaining number are largely non-union men, recruited from the open-shop mines of West Virginia and other states south of the Ohio River and are equally inefficient." The union in Nanty Glo also contributed stereotyped images of strikebreakers to its rank and file. A woman who was on the picket line at Heisley Coal Company gave me a hand bill that had been circulated by the union officials to picketers. It read:

A Prominent Clergyman once gave the following statement as his version of scabs, or strike-breakers, after having been compelled to associate with them a short time.

"After God finished the rattle snake, the toad and the vampire, he had some awful substance left, with which He made the scab.

"A scab is a two-legged animal with a corkscrew soul, a water-logged brain and a combination backbone made of jelly and glue. Where they should have their hearts he carries a tumor of rotten principle.

"When a scab comes down the street, honest men turn their backs, and the angels weep tears in heaven, and the devil shuts the gates of hell to keep him out. No man has a right to scab as long as there is a pool of water deep enough to drown his body in, or a rope long enough to hang his carcass with.

"Judas Iscariot was a gentleman compared with a scab, for after betraying his Master, he had enough character to hang himself—a scab has not.

"Esau sold his birthright for a mess of pottage.

"Judas Iscariot sold his Savior for 30 pieces of silver.

"Benedict Arnold sold his country for the promise of a commission in the English Army.

"The modern strike-breaker sells his birthright, his country, his wife, his children and his fellowmen for the unfilled promise of a trust corporation.

"Esau was a traitor to himself. Judas Iscariot was a traitor to his God. Benedict Arnold was a traitor to his country.

"A strike-breaker is a traitor to his God, his country, his family and himself and his class.

"A REAL MAN IS NEVER A STRIKEBREAKER."

The notion I described earlier of strikebreakers as animals is evident here and may have influenced the rank-and-file members. However, I did not hear any informant refer to strikebreakers as traitors. It seems as if the union rhetoric was likely to have fanned the already intense feelings of the struggling strikers. I spoke later with labor historian Melvyn Dubofsky, who told me that an antagonistic stance toward strikebreakers was the official stance of the UMWA since its inception.

The resentment stretched to include those who did not work during the strike but who nevertheless neglected their union duties with regard to union activities such as building barracks and picketing. These noncommitted union members were described by one man in terms that can best be described as freeloading: "There would be some people who earned their living in a mine but lived on land away from the mine. They didn't know what it was like to picket. They were rotters! We fought and picketed and went to court, and afterward they garnered in the profits." Resentment for this same characteristic was voiced with regard to those who came to mining towns after the 1927 strike. A miner's daughter told me: "There were ill feelings towards those who came here to work after the mines went union again." A former striker said likewise: "We were fightin' our asses off, and they'd come from Patton and get the best jobs."

The notion of non-union worker as freeloader was prevalent after the UMWA reorganized in District 2. A retired mineworker explained to me: "When the unions started back up again, it wasn't like today—a closed shop. You could work for the company and not belong to the union. There were some guys who stayed non-union all their lives. We used to say to them: 'You Scab!' It was around 1939. Sometimes the company would pay them a little extra to stay non-union. Some of those men stayed company men a *long* time." Today, the notion of "scab" has evolved to symbolize the freeloader almost exclusively. This is because the importation of strikebreakers has not been practiced in the mines of central Pennsylvania since the 1927 strike. However, non-union mines continue to operate during strikes. Union contracts, fought for by striking union members, are seen to increase the benefits of non-union workers as well as of the striking union members.

It is interesting to me that the wife of a miner from the 1927 strike combined these notions of thief and freeloader in her description of "scabs": "They just don't want to own up to what they did —takin' bread outa people's mouths. Like

today, the scabs get the same benefits as union men, and they
can't see it."

Actions taken against strikebreakers because of these feelings of contempt and resentment included shouting "scab" at
them, and crying "meeeoww" when a scab was seen sneaking
to or from work. These were meant to be shame-inducing.
Almost every informant remembered the picketers shouting
obscenities and cruelties at working miners, as well. Often
hostility erupted, and sometimes violence followed. At Nanty
Glo, one informant told me of a successful ruse to avert a violent confrontation: "Her husband had some words with
scabs from up on Pergin Hill. They came to his door and was
goin' to beat up on her husband. He told them: 'My neighbor, George, has a gun pointed at you right now!' He told a
lie to get them to go away." He also told of a case where the
threat of violence was very strong: "When the truck brought
the men home, the union men would be there waitin' and
would yell, 'You scab!' This one scab would come out on the
porch and sit on a wooden chair with his gun on his lap. I
remember seein' that as a kid." People told me of porches being dynamited at the houses of working miners in Spangler,
Winburne, and Nanty Glo. I heard of rocks being thrown at
strikebreakers in Bakerton, Spangler, Tipperary, and Nanty
Glo. A woman from Nanty Glo recalled the reaction to strikebreakers: "They first brought scabs into Heisley in open
trucks [past Webster Mine's houses], but the Webster people
hit them with chairs—threw chairs or whatever they could
find at the scabs. Then they started to bring them in what we
called paddy wagons. They was closed trucks."

A woman from Barnesboro recalled repeated instances of
gunfire at the home of a strikebreaker. Strikebreakers were
said to have been beaten in Dixonville, and a newspaper ar-
ticle (Barnesboro *Star* August 11, 1927) told of the homicidal
beating of a reputed "scab" in the town of Glen Campbell.

A feature of this animosity that I nearly overlooked was its
expression in school children. It was brought to my attention
by a man whose stepfather had returned to work for Peale,
Peacock, and Kerr in St. Benedict:

I tried to make friends with some of the kids. My father had started back at his own job in the mines. The kids said that he was scabbin': "Your old man's scabbin'!" Goin' ta school and comin' home I had two choices: run like hell or fight. I used to run—I was a fast runner. Sometimes they'd beat me up. After I got beat up a coupla times I fought back. Boy, I just fought like hell . . . I started to take my lunch so I didn't have to go home. If I had to go home [for lunch], I'da really been fightin'. I had to fight like hell as it was. I used to go home by [some of the kid's houses] . . . I'm explaining this to you so you would understand the animosity that there was at the time.

I heard reports by six informants of boys fighting in St. Benedict, four of these were unsolicited. Another example is: "There were new kids comin' in, and you always had to try 'em, and see how strong they was. There was this one Mexican kid who I wanted to try myself out on. The kids warned me that he'd pull a knife on me or somethin'. I waited until he was alone one night. I didn't really beat him up; I just showed him that I could handle him. [A kid I knew] was mean—he'd really tease 'em." Some of the stories were of strikebreaker sons beating sons of union loyalists. Cooper (1979a) has quoted an informant who recalled fighting between sons of strikers and sons of strikebreakers in Commodore. I also heard from informants of boys fighting in Winburne and in Marstellar. One informant from Nanty Glo told me: "It was exciting for us kids. We didn't understand—all we knew was that our parents told us we was union people and they was scabs, and that we was supposed to hate them. I remember that we'd sing: 'Scab, Scab sittin' in the grass, Union man go kick him in the ass.' and other rhymes that people would make up." It surprised me when he told me that he didn't remember fighting among the children in Nanty Glo: "In fact, sons of strikebreakers would be on the picket line with us!"

In St. Benedict, Peale, Peacock, and Kerr imported Mexican (or Mexican-American) strikebreakers. The Ku Klux Klan, which included some striking miners and at least one union officer, began to burn crosses again on the ball field in

St. Benedict, in an effort to intimidate them. The Klan had been particularly active in the early to mid-1920s. Cross-burnings had been prevalent in almost every mining town (Jenkins 1986). Despite the secrecy, most of the townspeople with whom I spoke knew the perpetrators. The postwar slow-down in the economy led to the conditions of labor surplus. The resulting competition for jobs and strikebreaking action led to a rekindling of prejudices against Catholic immigrants.

Nanty Glo (Singer 1982), St. Benedict, Bakerton, and Lilly (Johnstown *Tribune* April 9, 1924) were reported to have Klansmen in the union ranks, even among local union offic-ers, despite the threat of expulsion. My informants told me that most of the adult immigrants and many of the children were frightened by Klan marches and cross-burnings. How-ever, a man from Barnesboro and a woman from Nanty Glo told how, as adolescents, they fought back by burning circles on adjoining hillsides using a wagon wheel smeared with rail-road car axle grease. In Lilly, in 1924, a large rally and cross-burning was broken up by local teenage boys. A riot ensued in which three people were killed and 19 were injured (Johnstown *Tribune* April 7, 1924).

What was the effect of insults, violence, and Klan activity on strikebreakers and their families? One man recalled his fa-ther's fear of violence in crossing the picket line: "My father was a pumper in the mines—Barnes's in Philipsburg. He had to go to the mine to keep the mine from flooding. The union men pulled a gun on him, so he had to go to the mine with a revolver. He used to take me in the mines with him. He thought: 'If someone is with me, they won't shoot me.' My father tried to explain to them that if he didn't pump the mine it would flood, and then they wouldn't have any place to work after the strike." According to family members of striking miners, the strikebreakers were intimidated by the insults and violence. Some were afraid to go to work. Fami-lies were isolated on company property—afraid to venture into towns like Gallitzin or Nanty Glo: "The companies fi-nally broke the union, and everybody worked without a union. A lot of scabs left here. They got outa here. They couldn't stand the heat. They felt isolated here in the company houses—they couldn't go to town."

Even non-union workers in large non-union company towns like Colver felt isolated in their towns due to the hostile atmosphere outside. A Colver resident told me: "If you went to Nanty Glo you got murdered—you was a scab! You'd get throwed outta town if you went visitin' your father or brother. It happened to me once at a little mine right offa Spangler." Not only fear, but shame isolated strikebreakers: "People would go to a nearby town to work, for example Revloc. They were ashamed to stay and work as non-union workers at the same place. They would be considered scabs." Some men were motivated to join (or rejoin) the strikers: "After a while there was scabbin' goin' on—they had to. My husband worked one month, and they throwed stones, and he said, 'I'm not goin' back [scabbin']." The strikebreakers began to call strikers "rednecks," and in Marstellar the section of town where strikers lived off company property was named "Redneck Town." According to one informant: "The union guys and the scabs was always fightin'. Union guys used to call scabs 'cock suckers' and scabs would call union guys 'sap suckers.' One union guy prut near bit another guy's ear off!" The contempt-filled and resentment-filled hostilities did not end with the breaking of the unions, nor with new union organizing. For years, loud arguments erupted at picnics and in bars:

Years afterwards the old union men and scabs would argue in the bar near where my husband and I lived [in Marstellar]. I wouldn't allow my children under that front porch to play because of the words those men were usin'.

In church, too. [In Gallitzin] one striker saw the superintendent at a Catholic Church service. He said: 'If that son-of-a-bitch is goin' to be in this procession, I'm leavin' ', and he walked out of church. There were a lot of bitter feelings—and there's a lot of it still hangin' on today! In bars you'll still hear one man say to another, 'You're nothing but a 1927 company hero!'

I can remember some years later that there was a big picnic up on the ballfield [in St. Benedict] for some occasion. Some people who had left during the strike came back. If you had heard the fighting! They brought it up that some of the people had scabbed.

There are good men and there are pricks. The local wanted us each to chip in a dollar for burial of one of those men who broke the union. Why should I? He robbed from the men. He was against everything the union stood for.

Many of my informants could remember the names of men who worked during the strike and still lived in the area; some were eager to tell me the names. One woman's sentiments would probably be agreed on by the rest: "funny how you never forget." Several union men who still harbor ill feelings toward those who worked told of how they did not let actions betray their feelings:

I'll tell you, I know someone who I talk to quite often. But to this day when I see him I think to myself: "scab." That shows you how long these things last.

There would be a thousand of us on the picket line. It would stretch from the railroad crossing by Watkins to Sterling No. 1 [in Bakerton]. Yet some people who lived on farms would go back to work—maintenance they called it. When I think of it I get so *darned* mad! There was a whole family of 'em in Carrolltown. When I think of it I want to B-I-T-C-H, but I don't.

There's the guys who broke the union. I know who they are, but I don't hold grudges.

Here the emotional practices of resentment and contempt include talking to others about the morally questionable behavior but managing their words and demeanor in certain contexts, such as when speaking directly to the perpetrator. This would involve impression management of the type suggested by Goffman (1959). Hochschild (1983) has suggested that such "surface acting" involves a form of labor. Taking this a step further, I can understand how social interaction subsequent to the 1927 strike, and even peace-making between factions, as described below, involved much labor—an added burden of physical exertion beyond that required in the mine itself.

There were three instances of informants mentioning some resolution of the emotional hostilities:

After things quieted down we found that some of the scabs were the nicest people. They were just trying to make a living. It felt like they were taking bread out of your mouth at the time. We were all fightin' for survival.

A few scabs stayed [in Dixonville]. Some turned out to be as good a union guy as anyone. The majority left—they had to.

Some pretty nice [scab] families were left after the strike, whereas they were tearin' out each other's hair a year before . . . Some of the [West Virginia] scabs worked themselves in—they wanted to be like Pennsylvania miners. It was the women who brought them together.

Two accounts suggest that for a few strikers, resentment and contempt were not displayed, even though they were in the same social milieu as the other strikers. Two men told me of trying to use reasoning or persuasion to deter strikebreakers from entering the mine. A man from Clymer told me: "I remember picketin' with four other men at Garman, near Barnesboro. We talked peacefully to some miners headin' to work or asked them if they could help us. We ended up bein' invited in for breakfast. I don't know if they stayed out then or not. They did *that* day!" And a former miner from Nanty Glo said: "I used to stand on the picket line and try to reason with the men goin' in. I made friends with hundreds of scabs in the town, but I never scabbed a day in my life." In this second statement there is evidence that the speaker still sees "scabbin' " as an act lacking moral integrity, though he is not using it to chastise others.

Blacks were not used to any extent as strikebreakers in Central Pennsylvania during the 1927 strike, although some informants remember them being employed in Colver and Revloc during the 1922 strike. Noting the long-lingering

resentment of union men for strikebreakers, it seems possible
to track the origins of racial bigotry in those areas of Pennsylvania where black strikebreakers were more commonly
employed.

Those facing the decision of working or remaining on
strike were plagued by emotional conflict: on one hand, they
felt the shame of not being good providers for their families,
the restlessness associated with performing no physical activity, the shame of disloyalty to the operator, and the deprivation associated with barely getting by; this was countered by
the pride in union loyalty. On the other hand, to return to
work meant feeling the shame of disloyalty to the cause, a
feeling of guilt at depriving others of jobs and of life's essen-
tials. Even the men who considered going back to Sterling
Mines when they reopened on a non-union basis in December 1927 had great emotional difficulty. An informant told
me:

Well, people needed a bite to eat—that's what they'd say, "I need a
bite to eat so I gotta go scabbin'." I seen one brother hangin' on another brother cryin': "My family has nothin' to eat or wear, and winter's comin' on. So I have to go back!" His brother, who was doin'
better, would decide to stay out ... I actually seen this—one
brother cryin' to another brother about needin' a bite to eat. It
would happen—you'd see one of them cryin'.

Most of my efforts to meet with former strikebreakers
from the 1927 strike were unsuccessful. The men did not
want to talk to me. One company man, who by several independent reports had worked during the strike, told me that
he was living in a town thirty miles away at the time. I talked
to one former mineworker, the son of a farmer, whom I later
found out was enlisted to work in the mines during the strike.
I suspected his status when he claimed that no one was
evicted from company housing in 1927—I had more than a
dozen reports to the contrary. He claimed that a lot of people
anticipated that the operator would shut down the mine, and

"I seen one brother hangin' on another brother cryin', 'My family has nothin' to eat or wear, and winter's comin' on. So I have to go back.'"

on that account left. He had been working as a delivery man for the company prior to 1927 but took the position of dumper [tipple man] because the operator told him personally that he had nobody to dump coal "that could stand to work."

So most former strikebreakers were reluctant to talk to me; at least one appeared to lie about his whereabouts at that time. The one former strikebreaker who did talk to me described the company in favorable terms and the union miners in questionable terms (they didn't want to work and they abandoned the company). It would seem that fear of contempt from others still exists in today's context.

I also talked with a man from near Clymer who did not work during the strike but who did not participate in union activities. He told me, "They kept the mines workin' during

the 1927 strike. I was off for fifteen and a half months."
When I asked him if he picketed he said, "I didn't bother nobody." When I asked if he was evicted from housing he said,
"We had our own house and farm." Presumably, men such as
this felt no resentment or contempt for either camp. He
could always rely on the security of the family farm.

Contempt and resentment were not only held for those who
worked; it was expressed for the "pussyfoots" as well. Both
feelings related to policemen's unconscionable acts of striking, chasing, and trampling unarmed men, women, and children, of treating strikers and their families with a lack of
respect.

For most people with whom I spoke the term "pussyfoot"
was applied indiscriminately to any of the "outside" policemen who exerted force at the time. Most often they were talking about the Coal and Iron Police. These were state-
authorized, uniformed, and untrained policemen who were
paid by the coal companies to deal with strikers. Other police
officers included the state constabulary, state police, and special deputies. After the Coal and Iron Police were abolished
by Governor Gifford Pinchot, the companies hired their own
police. These were also referred to as pussyfoots. People had
various notions of who these men were: hoodlums from the
streets of Philadelphia; hoods the companies picked up in the
cities; unemployed men from Chicago, Cleveland, and New
York; a bunch of criminals on probation. Residents all agreed
that most pussyfoots were mounted on horses. In St. Benedict and Marstellar some rode in open-top touring cars. They
were always armed with pistols and night sticks with lead tips
for extra force in striking. Several residents remembered
their use of whips, and one informant told me: "They had
extra-long reins on the horse so they could use them like a
whip to wail on you." One former resident of St. Benedict recalled how they were outfitted there: "The Coal and Iron Police had .38 special revolvers, 30.30 rifles with a band of
30.30 shells across their shoulder, and sawed-off shotguns."

Their uniforms were described to me by a Bakerton native: "The Coal and Iron Police wore wrap-leggin's like in World War I. Some companies had khaki uniforms; some had brown, whatever. They was all deputized. The State Police had helmets, like a British policeman would wear, and their uniforms were real dark gray or black." Occasionally, strikers acted on their resentment and contempt. In St. Benedict the policemen were said to have been fired on as they sat in their guard shanties. Women were said to have shouted names and catcalls at them in Barnesboro, St. Benedict, and Spangler. One Nanty Glo resident was restrained from acting by relatives:

We were on a picket line and the pussyfoots come in on horses swingin' night sticks. They hit my father over the head. He told [the pussyfoot], "You wait here, I'll be back in a half-hour." One of the other pussyfoots told the other, "You hit the wrong man this time. He's goin' for a gun. If I was you, I'd get outta this area. He doesn't have a job; he doesn't have nothin' to lose." My dad went for his shotgun. We had to run home and sit on him to stop him from gettin' outta the house. He woulda shot that pussyfoot.

Contempt was expressed for a former Coal and Iron Police Officer who continued to live in Colver:

I can still remember the pussyfoots livin' in Colver. One of them went blind and had to have a seeing-eye dog. He'd be sittin' on a stoop, and people would walk by and spit on him. That sounds awful, but that's how bad the people hated him. Another one when he died had to be sent out of town to be buried. No undertaker in the area would take him.

Resentment was expressed for a former Coal and Iron Police officer who continued to work for Bethlehem Mines Corporation after the police force was dissolved:

When the unions came in, the company put one of [the Coal and Iron policemen who beat my brother] in as foreman over outside

work. I'd see him and tell him: "You try that today, you S.O.B.—
I'd shoot you right between the eyes." He wouldn't look at me. If
he saw me comin' he'd walk the other way. I heard that he was injured when a truck backed into him—probably someone gettin'
back at him.

To these feelings of contempt and resentment for the company police were added feelings of fear. Although many children were brave enough to shout "pussyfoot" at them, most
children were afraid of the mounted and uniformed officers.
As a girl, a woman informant had seen a mounted policeman
ride his horse onto the porch of a hotel in North Barnesboro
to chase a woman heckler. She described her feelings to me:
"I felt frightened, of course. Deep down I felt, 'If they're capable of that, they're capable of anything.' I also felt excitement." A man who was a child in 1927 said: "At night you
wouldn't dare go out. As a kid I remember hearing yelling
around and commotion—the galloping of hoofs of horses.
Mom and Dad wouldn't let us out." A man from Bakerton described the influence on children:

Once a woman had a hard time tryin' to get her eleven- or twelve-
year-old son to go to the company store for her. "That pussyfoot
with the big stick is down there. I'm not goin' there." Yes sir! you
couldn't get a child to run an errand for you. There was always a
pussyfoot hangin' around the company store.

A man who was exposed to Bethlehem Mines' Coal and Iron
Police had this to say:

When we saw a policeman we saw Hell! We would run and hide under the bed. We knew what they could do. We had fear. If there's a
hell, they're burning in it. No [com]passion, no nothin'! They hate
for no reason! It was a feather in their cap—the more dirt they did,
the more they were recognized.

He also told me:

You never forget these things—they stick with you! It was brutality!
They stick with me because I saw it. It was hard to believe how one

individual could have that kind of power. It was the same as the Gestapo . . . You'd think that you were goin' to grow up to be nothing.
There was no light up ahead. Always you had that fear . . . We
didn't know if our parents would be killed or maimed or what.
Where would we go for help? There was no help!

Not only children, but adults, were fearful. A Heilwood resident told me about the Bethlehem Mines' Police: "You could
be beaten to Hell and back! They were deputized by the
county sheriff. They didn't hesitate to hit you over the head.
They beat Charlie Bollag." I asked if there was any retaliation
by the miners. "No," he said, "people was afraid of them!
They didn't stand a snowball's chance in Hell!" Another man
told me of his recollection of the police: "At night the pussyfoots were nothin' more than marauders, tryin' ta scare out
the non-evicted people." Stories by the strikers suggest that
the police in some towns became fearful of the miners:

There was a pussyfoot shanty down on the road, but they were seldom in it. They were too afraid of gettin' beat up.

I remember once my uncle was in the basement shavin'. My father
was standin' on a bank beside the house. There was a scuffle with
the Coal and Iron Police below. My uncle ran out with a razor in
his hand. That Coal and Iron Policeman turned his horse around
and run!

The Coal and Iron Police got shot at in the shanties at No. 9
and No. 10. People on the hill with high-powered rifles. They got
down on the floor and crawled out the back. They said, "The Hell
with 'em! I wasn't goin' out after 'em when they're shootin' at me
at night!"

Despite a bitter strike in 1922 and company breaking of
union agreements in 1925, the great strike of 1927 came as a
shock to the union, the miners, and their families. In fact, a
former resident of Emeigh Run described the strike as being
"like a slap in the face," shocking and sobering as people were
forced to leave the area for an uncertain future. It seems possible to me that strikers may have had early expectations of

winning the wage increase. After all, the strikes of 1919 and
1922 had resulted in increased wages. Even though the 1922
strike lasted six months, operators with union contracts did
not resort to tactics such as strikebreaking, evictions, and
Coal and Iron Police. At least thirteen of my informants mentioned the 1922 strike. They may have had faith in the Central Pennsylvania operators, particularly Rembrandt Peale,
who had always chosen negotiation over exertion of overt
power. Several operators had expressed an interest in avert-
ing a strike and had offered to continue paying the current
Jacksonville wage following the April 1 deadline (UMWA 1927).

For most, the period of deprivation lasted from 1927 to 1942.
For those who were involved in union-breaking actions in
1925 and 1922 the duration was even longer. Although most
informants denied feeling deprived because "nobody had any
money," their behaviors of earning as much money as possible and of providing their children with lives free of want
may suggest otherwise. One man admitted feeling that he
starved during his father's two-year absence from work:

In 1928 a baby was born. That made us a family of ten people. We
was starvin'. I don't know how we existed—through the charity and
kindness of neighbors. I know that my dad went deeply in debt. He
was payin' it off for years. Nobody knows what hunger is until
you've been starvin'. If it wasn't for the garden and stored vegetables we wouldn'ta made it. I'm not ashamed to tell anybody this:
during that strike we had a garden in Watkins where the heads of
cabbage grew up without the bottom leaves. They were picked and
cooked!

A woman remembered sacrifices made by mothers:

People just didn't have anything. I still wonder how and what we
ate. I didn't know that mothers ate anything but coffee soup. His
[her husband's] mother and mine ate last and ate what was left.
Usually that wasn't much.

Still another resident remembered going hungry:

> There were ten years of bad time. As a lad I can remember going hungry. For breakfast we had coffee soup—some milk and bread in coffee. We went on that practically all day. Sometimes we'd have to repeat it for the second meal. We always managed to have bread.

At least ten people mentioned feeling that their lives were special because of having experienced the long duration of hard times:

> We, who lived through the depression as children, have a holiness— a respect for even a penny. The best part of those days was a spirit of togetherness. They don't have it now. Everybody helped everybody . . . It was a happy town—it wasn't deprived. We weren't under any stress or strain—nobody had anything.

> In the good old days things was rough, but people was like family. The country lost something they'll never get back . . . Anybody who's seventy or older and if they come from mining families they seen tough times—but it made them like iron.

> What we experienced makes you a better person, and you value what you've got.

> We went through the rough, but today I think it was educational. I have something in my life that other people don't have. It's good to reminisce.

Thus it was seen as a strengthening, a type of grace, an enriching experience, a lack of stress, and a unifying experience.

The dislocation of families caused anxiety and sadness in both the evicted families and in those of the townspeople who stayed. Those who had to move missed their homes and friends. One woman mentioned that it broke up her family as older siblings left the area. People were anxious about an uncertain future. As one woman put it: "The biggest part of the people moved out during the strike. A lot came back before the strike was ended. People, then, didn't have a dollar to move with and didn't know where they were going." Discouragement because of the length of the strike and the deplorable living conditions led many to leave the area and never return.

The families of management and of office personnel in St. Benedict mentioned how the character of the town had irreversibly changed. Besides this longing for the past, they expressed an anxiety about the roughness and transient quality of the strikebreakers moving in: "You can't imagine what it was like for a nice town to see people leaving and a rougher group moving in. The people who stayed hated to see it happen. It was a complete turnaround. Those who left were really interested in working and in having a nice town. The strikebreakers coming in were only interested in money."

The informants I interviewed were either adolescents or young adults at the time of the strike. The sons and daughters of strikers expressed an admiration for their parents' generation—for their loyalty to the union and for its courage in dealing with police and eviction:

The old guys—I praise them. But they're not around to enjoy the benefits. Some never got a pension or social security.

I seen Caleb talkin' eyeball-to-eyeball to the Coal and Iron Police. They was puttin' them outta the house, ya know. I was just a boy [eleven years old], and that really impressed me.

People were run out—evicted—families with eight to ten kids. They'd live in shanties down at No. 10 and at the barracks that the UMW built in Foxburg. They were good union men. They tried to save the union, but the companies were too powerful.

Children from the 1920s and 1930s expressed admiration for their parents' resourcefulness and hard work during the economic hard times. Union and non-union families alike suffered during these times. One such interviewer told me, "Dad always lived at a place where he could also farm. My mother was a hard worker—she did the farm work." Another said, "Pop was a good provider—we always had a cow. Mom was too. Mom was a wise old woman."

In what way might we better understand these emotions of a turbulent time? It is convenient for me to sit, more than sixty years later, in my relatively secure academic position and question the practices and worldview of destitute workers who were powerless to control many of the events occurring in their lives. My purpose is not to judge mining community members but to understand them in their relations with others within their class and with those in power.

I felt in listening to accounts of the last great strike that contempt and resentment represented favored social practices or roles in the drama. I recognized these as especially important by means of their number, types of, and frequency of culturally acceptable emotional roles people played, rich in discourse and based on context. This makes more sense to me than the distinction Levy (1984) drew between hypo- and hypercognized emotions. With this distinction, Levy emphasized linguistic factors such as the number of categories for one emotion as well as its acknowledged cause and evaluation. It seems as if he emphasized emotion discourse (i.e., statements *about* emotions) to the exclusion of emotional discourse. My analysis emphasizes the latter. Abu-Lughod and Lutz (1990) have recently drawn this distinction in types of discourse.

I find that for working-class Americans, and perhaps for most Americans, the naming or identifying of their emotions is seldom practiced in everyday discourse. For example, "I am contemptuous of you," or even "I resent your behavior," would not likely occur in emotional conversation. An outside observer might be able to label the emotions as contempt and resentment based on context, however. In this case, group members have not learned emotion terms but rather phrases and body *hexis* (Bourdieu 1984) to accompany it, e.g., a contemptuous sneer, or the phrase "How dare you. . ."

As Sarbin (1986) pointed out, the notion of dramatistic roles does not deny logic to the emotional encounters, but this logic may have been learned, along with values and identity, from childhood exposures to similar fictional (and I would add real) dramas. The mining family children and young adults witnessed demeanor, behavior, and discourse. They learned, especially via the rhymes, derogatory descriptions, and rhetoric, how to respond emotionally to strikebreakers and to company-paid police.

So, the contemptuous sneer, the tone of voice, the acts of describing others as animals, spitting at others and making deprecatory remarks were all acceptable practices within the system of the *habitus*. These were expressed as acceptable practices with regard to certain categories of workers, i.e., "scabs" and "pussyfoots." According to the thinking of Bourdieu, the community, through these practices and discourse, intervenes between the child and the world. This is how the *habitus* is inculcated. Here also are the origins of the group *doxa,* i.e., their view of the self-evident or naturally appearing world. For example, the child may have learned that there are certain categories of workers, each displaying certain practices, each following self-evident motivations. However, with the *doxa* the limits to thought and perception are not recognized as such (Bourdieu 1976:164-68).

The miner's *doxa* viewed capitalist ownership as the only feasible system of work relations. The notions of worker ownership or government ownership of the mines seem to have been out of the realm of consideration. Either arrangement would have had the potential to eliminate the divisiveness within the workers' ranks. This arbitrary worldview, which is misrecognized as natural, was likewise reinforced by the discourses at both levels of UMWA leadership and by the coal operators. According to my informants, the coal operators reduced, through their discourse, the striking miners' actions to "not wanting to work." This exacerbated the moral antagonisms within the working class by giving the strikebreakers and company police a rationale for moral contempt of miners and their families.

One aspect of context I first overlooked was the informants' conversations with me. Here, emotional discourse was expressed, as well as remembered, within a new context. What type of relationship did we have? It seemed to me to be one in which the informants perceived me to be a friendly member of the mining community with empathy for their feelings. I was also a chronicler of events. As such, I represented someone who could record, adjudicate, and announce the injustices that had befallen them. I could also justify with a nod or by serious questions their feelings then and/or now concerning the events of those times. This was the pragmatics of the narrations, themselves emotional discourses—often discourses on discourse.

The resentment seems to have been an anger addressed to the behavior of the strikebreakers and police for which there was no compensation, whereas contempt was anger focused on the person. Contempt seemed to me to have several components. In one respect it displayed a derogatory image of the other—an image of being less than human. It also functioned as a type of moral indignation. As such it set boundaries within the mining class to mark out proper and improper behavior. Finally, it acted as a means of shaming the individuals; name calling and catcalls were meant to create discomfort in the body of the other. It was thus meant to punish him for his action and/or to change his behavior. Even the instance of a striker crying as he told his brother that he was going back to work because his family was starving may indicate the shame he anticipated bearing as a result of his decision. I also heard tales of men sneaking to and from the mines and, of course, working in the mines of other towns. Some men, such as the Nanty Glo strikebreaker who came out on his porch with a gun to face gathering strikers, seemed to be shameless with regard to this issue. So the social aspects of these emotions take on great importance eclipsing the psychological aspects.

I came to the idea while in the process of analyzing these emotional practices, that the emotional intensity of the 1927 strike misdirected emotional practices of workers away from the mine owners. The strikers saw themselves caught on the short end of power relations with regard to both strikebreakers and the Coal and Iron Police. They could see loss of security and of liberty, and they had few tactics with which to salvage them. They saw the "scabs" and "pussyfoots" as usurpers as much as they recognized the operators as such.

The miners' emotional practices, then, actually reinforced the power differences between owners and workers. It is difficult to say if the owners realized that their agents (i.e., "scabs" and "pussyfoots") would draw attention, and more importantly emotional practices, away from them. It may have been coincidental with the owners' concern with producing coal that could compete in price with the non-union-produced coal.

5

Emotions, Work Values, and Exploitation

Good times—hard times; that's a part of living. I always thought that a man was put on earth to work with his hands and to sweat.

—former handloader

What would coal miners do if the companies for whom they worked tried to change their job from independent artisanry to highly supervised crew work? This is the question Carter Goodrich (1925) asked himself as he envisioned the era of mechanized mining. Sixty-five years later I was asking the men who made the transition how they perceived the change. My understanding of how the men felt about the change differs greatly from the picture painted by Goodrich, and the miner's work ethic lies at the center of the discrepancy.

I became interested in the way coal was mined after reading Dix's 1977 treatise on the handloading era in coal mining prior to beginning my fieldwork. However, I did not realize the topic's importance to my study until two events had occurred during the early months of fieldwork. I had discovered in early interviews with retired mineworkers that there was a strong interest in talking about the techniques of handloading, which for most mines in the Central Pennsylvania bituminous coal district persisted until the late 1940s. The second event was my reading of Carter Goodrich's *The Miner's Freedom;* the concept of the freedom-loving rebellious miner intrigued me. Thus I began to ask retired mineworkers more questions about handloading and about the modern machine-loading techniques that replaced it.

Goodrich (1925) has described the coal miner of the hand-loading generation as an independent artisan bent at all costs on maintaining his freedom over production decisions and work schedule. Handloaders worked in teams of two, laboring in their own assigned work space using their own tools and explosives. They took down rock from the roof or floor for clearance for coal cars, laid track every day as they mined further, pushed cars by hand in and out from their working place, blew coal out of the seam using dynamite or black powder, picked out impurities from the loosened coal, and loaded coal into cars using shovels. If they had a wet spot they bailed water into cars they had sealed with old bags or clay. They cleaned up roof falls and loaded rock and impurities from their workplace into cars to be taken out of the mine or shoveled it to the side of their tunnel so that new tracks could be laid and a cutting machine brought in to cut the top or bottom of the coal seam in preparation for blasting the next day.

With all of this labor, the miners decided when and how things should be carried out since they saw the foreman only once a day. They could decide how much coal they wanted to load that day since they were paid by the ton for the amount of clean coal loaded. In other words, the handloaders made most of the production decisions.

Companies intensified their supervision and increased job specialization during the mechanization of the coal mines. These changes had taken place earlier in other industries, following the principles of scientific management put forth by Frederick W. Taylor (Dix 1977:60-61). This seems to follow the description of Foucault (1977), who wrote of a change in discipline from one in which there was an interest by the dominant in distributing bodies, extracting time, and accumulating it, to one in which there was a concerted effort to compose human forces in order to obtain an efficient machine. Goodrich (1925) foresaw problems in getting miners to accept a means of production involving a daily wage and the increased supervision required to coerce the men to maintain production.

Dix (1989), in maintaining Goodrich's image of the rebellious miner, has suggested that the miner's power was undermined by his own leadership. Dix put forth the well-documented argument that the Appalachian agreements that John L. Lewis negotiated, as well as Lewis's centralization of the UMWA's bargaining structure, promoted the introduction of mechanization and daily wage over the objections of the rank-and-file members. Interestingly, Dix reported that former miners who were informed of his conclusions disagreed with his assessment.

In the analysis that follows I will take readers on a journey toward understanding members of the hand-loading generation. The route I follow will necessarily involve several directions. I will analyze the responses of former handloaders to changes in production technique. I will analyze the emotional discourses of members of the handloading generation concerning certain workers in order to help define work values. And I will put forward my understanding of power relations between mineworkers of the handloading generation and management.

From my historical vantage point sixty-five years after the work of Goodrich, I could ask the men themselves how they responded to the new production techniques involving machine-loading, increased supervision, working crews, and hourly wage. I talked to 105 men from the pool of retired mineworkers who had made the transition from handloading to machine loading concerning their thoughts and feelings about the change in method of mining and mode of pay. I also talked to twenty-one former mine foremen and to more than one hundred other residents of the area. The most striking insights from these interviews are the apparent lack of recognition of handloading as a craft or a petty entrepreneurship, and the minimal concern with increased supervision.

Twenty-nine of the informants interviewed talked of the differences between handloading and machine-loading. I used no protocol or survey form to gather this data. Several responses included the mention of conveyor-loading. This was a transitional phase between hand- and machine-loading, in which men worked on teams of nine to nineteen men, shoveling coal from a kneeling position onto a conveyor that took it to cars. These men did nothing to prepare the coal for loading, but they had to help move the heavy conveyor equipment. The crew shared the tonnage wages.

Preference for type of loading was evenly distributed between the independent hand-loading and the specialized and supervised machine-loading. I found it interesting that earning more money and having an assured daily wage together were mentioned in more than half of the explanations for preference (seventeen times). Six of the miners who indicated a preference for machine-loading were interested in a guaranteed daily wage. Freedom was mentioned in one form or another four times.

As I interviewed these older informants, I was frustrated by the dearth of references to the handloader's freedom in their workplace. I began to ask direct questions about this topic, hoping that once the memory of this "luxury" was prompted a flow of relevant discourse would follow. Even with this imposition of my agenda into the interview, I was able to perceive a recognition of the meaning of handloader's freedom in only eight instances. In half of these cases the informants did not attach much significance to it. Most of the former handloaders had their own agenda to address in our interviews, and they held firmly to it. They seemed to be celebrating their work values. In reflecting back, most also seemed not to perceive any condition of autonomy during conditions of handloading.

A 1941 telegram from the District 2 president provided evidence of a threatened strike at Lucerne Mines over the installation of mobile loaders (Mark 1941); however, the cause of concern was not identified. Work freedom was not the obvious reason for the threat of strike.

I was told that some of the men who were older workers when machine-loading was introduced (and are now dead) were afraid of operating the machines and were consequently afraid of job loss. One informant told me that when machines came in at the Pennsylvania Coal and Coke Mine in Marstellar, the separate category of handloader was carried for less than a year. Then the handloaders were all laid off. Other than feelings concerning job displacement, there seemed to be no expression of rebellion by miners with regard to the introduction of machine mining and no remorse over having given up the shovel and explosives.

Archival sources provided other references to local miners' reactions to changes in the way coal was mined. A strike of 240 miners was reported at the Logan No. 4 Mine in Beaverdale. The six-week strike concerned the management decision to move Joy-loader crewmen to conveyor-loading. The move would involve a change from guaranteed high daily wage to a tonnage wage that was split among crew members. The strikers sought a guaranteed substantial daily wage (Johnstown *Tribune* August 1, 1945).

In a letter to the president of District 2 UMWA, a cutting machine operator complained of not being paid extra wages as a cutter on the conveyor team. The wages were divided among all crew members (Mark 1941). These pieces of information support the contention that higher wages and wage security were among the most important factors in the workplace for central Pennsylvania mineworkers.

In cases such as these where an expected reaction, i.e., rebellion, does not occur, the motivations of the groups involved aid in understanding the dynamics of power involved. Through the description and analysis that follow I suggest that, with respect to increasing management control over production, the miners' work values and attitudes overshadowed their concerns for production freedom.

By analyzing several types of discourse I found the characteristics of the older generation's work attitude to include: a respect for and loyalty to the company; an emphasis on earning as much as possible; a need for wage security; self-direction and responsibility; a sense of being a good provider for family; and a sense of just labor for an agreed upon wage. These values seem to have overridden miners' concern for freedom in making production decisions, thus making the companies' move to mechanize relatively smooth.

The work attitudes of these retired miners were expressed both directly (for example, in the process of expressing preference for the type of work) and indirectly (in commenting on the behavior of some young miners who labored in the 1970s and on the behavior of some miners following the institution of portal-to-portal pay in 1942). The latter comments are interesting because they reveal some meaning-giving perceptions of the hand-loading generation's *habitus* (Bourdieu 1984). Bourdieu described the *habitus* as "necessity internalized and converted into a disposition that generates meaningful practices and meaning-giving perceptions" (1984:170). Although Bourdieu did not write of work values or practices, these appear to have great meaning to members of the class of hand-loading miners. This would explain why members grew indignitant when they spoke to me about the discrepant work values and why they assumed that it was a discrepancy naturally understood by all.

The union contract for 1942 called for payment of mine-workers for their underground travel time to and from their places of work. This practice caused many of my older informants to criticize the actions of a small percentage of hand-loaders—those who could live on low pay. Because portal-to-portal pay provided a sizable base income for these men, they would load only one or two cars of coal a day. I found little support for this exercise of miners' freedom but rather a moral indignation toward miners who exercised freedom to limit their production and work time.

Many retired miners and miners' wives found this behavior to be ethically wrong, calling it freeloading, a lack of interest in working hard and doing an honest day's work:

Yeah, we had a lot of free-loaders here—like anywhere . . . For good wages you should get five or six cars apiece. Some guys were loading only two cars, and that didn't pay for portal-to-portal.

I never thought good about it . . . So many wouldn't work, and they would brag about it. I feel that you should do an honest day's work for an honest day's pay.

Some identified it as the cause for the end of handloading as a way of mining, and they considered it to be taking advantage of the company:

It should nevera been in there [the contract] if they was doin' piece work. I knowed guys would come in and load one car and go home. There were a lotta people gettin' money for nothin'—a lotta bachelors and boozers . . . They was takin' advantage of the company.

The mines had to go to mechanized loading just to keep up production. Men were gettin' paid extra for portal-to-portal and some weren't too interested in workin' hard.

Several characterized it as ruining the mines: "Portal-to-portal was the start of the downfall for coal mining."

I heard similar moral indignation voiced by retired mineworkers and their families for members of the younger generation of miners working in the 1970s. They saw many young mineworkers as not working hard:

Those younger men—none of them wanted to work. All they was interested in was how many years until their retirement. They weren't producin' anything. If I said anything they accused me of bein' a company man.

A lot of it has to do with the union: "It's not my job." A lot of 'em say, "I *put* my eight hours in," not: "I worked eight hours." Some men would leave if they got wet; some would deliberately get wet because they wanted to go home early. My husband used to say: "The older the man, the better."

Yeah, you couldn't tell [the younger guys] nothin'. One guy said he was goin' to get it [the mine] on strike—smart aleck; just a punk! Men were repairin' autos and jeeps in the shops. People spoiled it for themselves—they wouldn't listen. It wasn't the older ones. [The younger ones] would be leavin' early with all kinds of excuses Saturday night. They just ruined it. They had guaranteed pay, and they still wouldn't work—I can't understand it. They got it in their heads: "The company's got money—let 'em pay."

Like the residents who complained of the abuses of portal-to-portal pay, they felt that many younger workers were taking advantage of the company:

In the sixties and seventies the men were only for themselves. They weren't concerned with the company.

Another thing, they thought there was an unending level of finances . . . I told them: "You can only shake the tree so long, and all the fruit will be gone."

Some characterized the attitude as ruining the mines:

In the seventies some men would be quittin' early. That's the way guys was. I tried to tell 'em: "Boys, you're goin' to spoil this." The company had guys comin' in to check. They caught some of them there with one and a half hours of work time left. We tried to tell 'em, "Whatever you do keep on movin'." They didn't listen, and the company shut it down.

What ruined the mines was the wildcat strikes and the men not producin'.

I was struck by the similarity of these comments, although voiced by different informants, to those expressed with regard to portal-to-portal pay. The older people also complained of young men sleeping in the mines and bragging about it, about young miners not being self-directed in their work, and about workers refusing to work for reasons of safety or job description. These critical appraisals of workers indicate the values of the older generation. Such values were also revealed in positive statements about workers, particularly about older workers.

During my interviews with members of the older generation the reference to someone as a hard worker or as a good worker was voiced at least forty-three times. It was obviously a revered characteristic—one of which a person was likely to be proud. Two men in their forties independently expressed the observation that their father's generation would never give less than their best. One of these men described the attitude of the older generation miner as "There's no S.O.B.'s goin' to say that I didn't give 'em what they want." Several men indicated that they wished to please their boss and receive praise for their efforts. For example, one older informant stated: "Dad had a good job and the company liked him." A former handloader who left the area in the 1930s to work for a city contractor said: "My boss liked me so much that he put me on a service truck all by myself, doing service repair work."

There were seven references to working while injured, including two comments made by women concerning their husbands and one by a miner's widow who had returned to work in the dress factory only four months after open-heart surgery despite agonizing pain. Loyalty to the company and to the family were the most common explanations. However, one woman mentioned that her husband became restless when he wasn't working: "He couldn't not work." Several other people mentioned a restlessness when not doing physical labor, including an eighty-year-old retired mineworker with black lung who told me: "This livin' without workin' is no good."

There were also six references to moral obligation on the worker's part to help the company make money. One man expressed it as: "They're givin' us a living—we owe them loyalty." A miner's daughter who had retired from a local garment factory stressed the same sentiments: "He gave me a job so I felt I owed him to do the best I could do. By helping him I felt like I helped myself." Members of the handloading generation also spoke of their pride in work: "We always did good work, we made sure it was right. And we were never afraid of work; we always worked hard and fast. We had a system . . . We never argued who would do what; we just did it. There was no such thing as saying 'I did more than you did.' " They also mentioned the joy of work, saying, "When I was healthy, in good shape, I loved to work—whether in the mines or wherever it was. I'd go at it!" And "I enjoyed workin' in No. 10, handloading—I really did!"

After listening to my informants tell me of their work exploits, relationships, and values I found much to admire in the way they conducted their lives. Thus feeling fired up by the passion to defend my informants' values, I looked at Goodrich's analysis more critically. Since he relied heavily on archival material, perhaps his contact with actual miners was too meager to understand them. It is also possible that he, and later I, were projecting personal issues into the analysis. Academic freedom has always held high priority with university faculty; perhaps we thought that miners would likewise seek as much work freedom as possible. Bourdieu (1984:373-74) has addressed this sort of analytical distortion in the study of the French working class. He saw the intellectual as a member of a different *habitus,* comprehending the working class condition through schemes of perception and appreciation that are, in fact, foreign to members of the working class.

I was surprised to find clues indicating that, with regard to the reaction to the change in coal-loading technique, elements of the miners' work attitude were at least competing with the significance of what Goodrich spoke of as the "indiscipline of the mines." For example, he stated that a handloader's output was limited by factors out of his control—things such as number of available cars and thickness of coal seam (Goodrich 1925:32). This would suggest to me a frustration that might lead miners to embrace a guaranteed daily wage. In fact, such frustration and its association with accepting a position offering a daily wage were mentioned by sixteen of twenty-nine former handloaders who mentioned a preference for method. A majority of other handloaders also mentioned these conditions, including problems with water, with roof falls, with separating out impurities, and with coal inspectors. Those who preferred handloading did so because they were lucky enough to have worked where they were not

Cutting machines undercut the coal seam to facilitate removal by blasting. Goodrich called cutting machine operators the "aristocrats" of handloading mines. Courtesy of the U.S. Department of the Interior.

limited to any great extent by uncontrollable factors and were able to earn higher wages than day workers.

Goodrich (1925:45–47) mentioned the self-discipline of the machine runners who worked unsupervised to cut the coal for miners to shoot and load. He also mentioned that pay and pride of work made these machine operators the "aristocrats of the industry." He implied by this that pride, desire for pay, and self-discipline were missing in those who handloaded the coal. I found that the desire for increased pay and the pride in their work were often self-mentioned characteristics of the handloaders I interviewed. For them, machine-running was a desired daily-wage position, and they appeared to possess the self-discipline required for the job. How could Goodrich have ignored this certain contradiction in his analysis? Did he intend to imply that handloaders lacked self-discipline and a desire to be the "aristocrats of the industry"?

I found myself joining in the hand-loading generation's celebration of work ability and work attitude. I found these people likable and admirable in their conviction to make the best of daily toil. Yet, I still had to examine the issue of increased management force in the new practice of using workers as team components. It became clear to me that exploitation occurred in some way that was seldom, if ever, understood by the exploiter or the exploited. In this respect my perspective is similar to that of Gaventa, who, in his analysis of power and powerlessness, described a "third dimension of power," which allows for both the dominating and the dominated to be oblivious of exploitation and accepting of the *status quo* (1980:11).

Hochschild (1983) has noted how emotion can be manipulated in self and others to achieve the ends of American corporations. In her analysis, the emotional manipulation is self-inflicted by the worker as a means of interacting in a specific way with customers. The technique was different in the Pennsylvania coal mines where managers exploited the emotional dispositions or values of mineworkers. The fundamental attachment of values to emotions, as suggested by Middleton (1989:198) and Gerber (1985:151) seem to provide the clue to understanding this process. Gerber succinctly described the process: "If the feeling itself is defined as the proper reaction of a 'good' person, and if the behaviors consequent upon it are socially valuable, an emotional disposition toward socially correct action is created . . . The emotion therefore reinforces important social values. If the feelings in question rest on a basic affect, the emotional force of that powerful underlying psychological pattern is therefore put to the service of society." Despite the fact that she wrote here in terms of internal feeling states and underlying psychological patterns, Gerber seems to have been relating them to emotional practices and dispositions. In this respect her ideas resemble those of Bourdieu (1976, 1984), who wrote of dispositions and practices as being a part of a culture's or a class's *habitus*. Her notions also seem compatable with those of Abu-Lughod and Lutz (1991), who viewed emotional practices as the most significant aspects of emotions. Using the perspectives of these authors I would define the work ethic as a set of learned and lived emotional practices.

The change in the quality of labor was made by taking advantage of the men's desire to please, of their fearful expectation of bad economic times, and of their moral pride in meeting the stipulations of an agreement. In addition, the foremen could increase control of the men by inducing shame or the fear of being shamed in the mines. In the remainder of this chapter I will describe in detail the nature of these emotional dispositions with regard to work. I will begin by looking at loyalty to the company or the foreman. Many of the men with whom I spoke made a mental distinction between company and management. For some, respect for the company or owner often coexisted with a lack of respect for company management. One former miner, in speaking to me about safety, said: "A company is an intangible product. It furnishes equipment and hires men to run it legally. So many do it illegally—how can you blame them? [Safety] has to be between the management and the men." [with regard to safety] In another instance, a retired miner expressed his opinion with regard to the Valdez oil spill in Alaska: "It wasn't Exxon's fault; it was the pilot's."

Less commonly, men would express praise for the foreman and anger toward the company, the owner, or the superintendent. This mental splitting of company from management was often tied to a thankfulness for being given the job, and it allowed for a dedication to work responsibilities almost regardless of the work situation. One former mineworker summed up the sentiments well when he told me: "Most of these younger guys don't know that you can disagree [with the company or management] and still give them all you got." This loyalty was not likely to be held by defeated union workers during the time without union (1927 to 1933) but might have been expressed after the union had been reinstated.

I became fascinated by this sense of loyalty and by the overriding importance of increased earnings. Through the

accounts of my informants I was able to piece together their
sense of why these factors were so important to their work
lives. In some cases there appeared to be a strong emotion in
the desire to please. A co-occurant factor was the set of expectations arising out of their socialization during times of
persistent deprivation. I will detail below my understanding
of these processes.

The desire of miners' children to please their parents was
suggested to me by the statement of an older woman who was
the daughter of a former handloader and the wife of a mineworker: "The ethnic traditions are the source of the work
ethic. It was a source of pride. It was how they could please
their parents. It was the ultimate identification of the child."
The desire to please was carried into adulthood by at least
some of the former miners. This reminds me of the psychological phenomenon that Arieti (1972) has labeled "endocracy." It involves a sense of "I ought to" arising out of a need
to please parents. The child adopts patterns of behavior he
believes the adult wants to see reproduced.

It is easy for me to imagine that parents would utilize their
children's eagerness to please as a way of enculturating the
values of hard physical work, providing for the family, and
producing the most fruit from their labor. It may well have
been an unconscious process. As adults, people could then
transfer the need to please parents to authority figures who
praise those same values, particularly if their role models displayed similar dispositions. The disposition to please might
be demonstrated later when as an adult the miner expressed
a loyalty to the company.

Retired handloaders and their kin informed me that the
handloading system encouraged the working together of
fathers and sons. In order to get mining papers, a novice
needed to find someone to work with for two years. In addition, a novice often needed someone to "take him in" as a
means of being hired. Virtually every handloader with whom
I spoke had been taken into the mines by his father where
they worked side-by-side for more than the two required

years. Most of the entering miners were sixteen years old, but many had had prior experience working with their fathers in small country bank mines or during "off-shift" hours in the big mine.

John Brophy (1955:95), a UMWA district president who spent many years as a handloader in central Pennsylvania, indicated that to be "taken in" by one's father was of great emotional significance: "I got a thrill at the thought of having an opportunity to go and work in the mine, to go and work alongside of my father . . . I was conscious of the fact that my father was a good workman; that he had a pride in his calling." This sentiment was suggested also by the responses I obtained from men who had lost their fathers and thus the opportunity to be taken in. One man told me in a sad tone: "The other boys had their fathers to take them into the mine, but not me." Another said: "One of [the boys] was tryin' to calm me down, and I punched him in the nose. They all had fathers to take 'em into the mine to get good jobs." The son of a former handloader told me that one of his favorite accounts was his father's description of working with the older man who took him into the mines and of the feelings involved: "I remember my dad talkin' about that. His dad died before he was old enough to go into the mines. He went in with an older man. He had a vivid memory of it before he went to the hospital. He had a spell when he talked to us for three to four hours as if he was living forty years ago. He talked a lot about when he first started in the mines and his strong feelings for the older man and about what a good man he was."

Production teams consisted of two men who shared the duties of mining their "room." This extensive working with one's father or with an admired older man could conceivably enhance the chance of endocracy being important in the adult lives of the miners. One man in his mid-forties who considered himself one of the last to be indoctrinated in the old ways spoke to me about the indoctrination by an older miner assigned to him for one year. This was in a mechanized mine of the early 1960s.

I came in under the old system and four years later I was foreman under the new system. So I saw both sides. I really respect being indoctrinated in the old ways—first by my father and then in the mines . . . I was indoctrinated . . . to help the company. You don't wait for a mechanic unless it's serious. You do it yourself or take off panels [from the machine] after you make a call for the mechanic. The older guys were hard workin' people. They had pride. They taught me values in terms of pride in workmanship . . . They taught you "Don't waste material—it costs." . . . Undoubtedly there was more respect for older men when you were brought in. I remember, I was timbering for the man who brought me in—he was runnin' a miner [operating a continuous mining machine]. There was a draw-slate fall [roof cave-in], and it hit the roof of the miner and then the right side. He said to me: "Kid, I think you saved my life." I'll never forget that.

The notion of endocracy could also apply to the work values of self-direction, emphasis on earnings, providing for family, and working hard. When the father and son labored together both paychecks went to the family, which usually meant to the novice's mother. The same was true for unmarried daughters who worked in factories or in other's homes. Here, again, was a situation associated with trying to please one's mother. Later, the desire to earn more money could satisfy the need to please.

Working alongside a father or older man taught a young miner to anticipate what would be needed to keep them working efficiently. Again, eagerness to please could explain a self-directed worker, i.e. one who knows what to do without being told.

Statements by former handloaders and foremen alike suggest that foremen made it a practice of hiring the sons of hard-working employees with the idea that the son would display a similar work disposition. Parents were likely to realize that this apprenticeship also represented an evaluation period. One miner's widow told me: "If you're a good worker, they take you." It is questionable as to whether the coal operators had any sense that a novice's indoctrination into mining might foster endocracy. For them, it was more economical than having foremen teach new employees. The industrious, loyal, and efficient handloaders may have been the unrecognized bonus.

When the mines introduced conveyor-loading, most of the miners gladly gave up the buddy system and making production decisions because they were loyal employees who saw a chance to load more coal, to make more money, and to better provide for their families. The fully mechanized mine was thus seen as one more opportunity to meet those same goals.

The aspirations to earn more money could also be tied to the person's expectations. Gutman (1976) has suggested that aspirations and expectations can help to shape behavior. Workers whose childhoods and early adult life were characterized by years of deprivation because of major strikes, unemployment and slack employment might live with expectations and fear of the return of the long "slack time" of 1922 to 1942. Feelings from times of deprivation persist among members of the older generation today. One woman told me of her security related to canning: "I like to just look at our canned goods sitting on the shelf. For awhile I was afraid we wouldn't have anything to can because of the drought." Another woman told me an adage that could only relate to the most desperate of times: "They say that a good mother will feed twelve children with one egg." Most of those who had experienced the persistent depression in the area were eager to tell me how hard the times were and how they coped.

Later loyalty to company and a desire for increased earnings could be directed to some extent by the fear of poor times to come. This fear of deprivation was one emotion that some companies manipulated to their benefit through the firings and threats of firings that were reported to be common in some mines prior to and during the era of mine mechanization. It is interesting to note that Barbash (1983) has suggested the possibility that the work ethic of older-generation Americans sprang from the era of scarcity and deprivation.

Pride in one's physical abilities at work meant much to these men. Many told me of the number of tons they could load, the neatness of their workplace, their adeptness with the cutting machine, or the amount of money they could earn in comparison to others. They had been enculturated to be physical doers and were proud of their physical accomplishments. They were undoubtedly warmed by the praise of younger siblings, peers, parents, older workers, and even bosses. If harshly criticized or rebuked such men might feel humiliation or shame. This is an important point since the inducement of shame as a social practice could be utilized in power relations. When referring to shame, I am using Kaufman's definition: a feeling of being "seen in a painfully diminished sense." It is a "piercing awareness of ourselves as fundamentally deficient in some vital way as a human being" (Kaufman 1983:8).

Pride and shame were manipulated by the management's system of "knocking" as described by Goodrich (1925:50-51). The system was used during handloading days as a means to hold day workers such as track layers and motor men (underground locomotive drivers) to a disciplined work schedule with little supervision. Men would be "knocked," i.e. blamed and/or shamed by the foreman or by coworkers, if they did not do suitable work. For example, the foreman would "knock" the motorman if the coal didn't come out of the mine fast enough; that man would, in turn, "knock" the tracklayer if a bad rail had wrecked his cars. Likewise, the handloader would get the boss to "knock" the driver if the cars were slow, or the tracklayer, if a new switch had not been laid.

In an effort to test Goodrich's assertion, I asked a particularly knowledgeable informant if he had ever heard of "knocking." He replied: "Yes, I heard of it. They also called it 'biting'. Somebody had to get the blame." In this system of "biting" it was customary to avoid or lessen the shame associated with being blamed for poor workmanship by transferring the blame. Thus blame and shame could be distributed throughout the workforce to control work effort in the absence of direct supervision. This system kept men working in an effort not to be shamed. Brophy (1957:98) spoke of similar self-management through shaming among the handloaders themselves: "Also, unless there are very good reasons, such as adverse conditions developing in the workplace, the handloader is inclined to lose face with his fellow workers if he misses his turn [of cars] because of poor workmanship."

The practice of control through "knocking" was continued for some time by foremen of the new mining technique. Older miners who had begun working during the handloading era seemed especially vulnerable to control by being "bawled out" and blamed for their inadequacies in front of their fellow crew members.

Men told me numerous anecdotes indicating that the emotional manipulation that originated with knocking still existed in the modern mines. For example, a retired foreman told me: "I told those guys, 'Why don't you go backward—not forward—when you pick up your check!' God-damned guys didn't want to work!" In effect he was telling his men that they should hide their face when accepting money for such poor work—that they should be ashamed. It was not just mine management that used emotional manipulation in this region. I was told of a calculated strategy being used currently by a regional garment factory. It was relayed to me by a retired garment worker who was the daughter of a handloader. "They had to pay at least thirty-five dollars per day [to piece-workers], but if you were below that they put it in red on your pay stub. Nobody wanted red on their stub. I'da been humiliated. They wanted to let you know when they were pleased with you." Here sentiments of shame and desire to please were manipulated.

Another value related to hard work was the older miners' sense of respect in honoring the terms of a contract by working hard. In all probability it arose from sources such as earlier generations' peasant and artisanal values and from early union notions of a just wage for a just day's work, as well as from religious morals regarding honesty. A former handloader working in a mechanized mine in 1969 spoke of the moral nature of working hard: "But the miners weren't fair

either—standing around doing nothing. That's outright
stealing. My brothers were all the same as me. We would always put in a hard day's work. The men would say that the
company was makin' millions. I don't care if they make millions—that's why they're in business. That's no reason for me
to shirk my duty." Yankelovitch (1974) has pointed out a tie
between the religious and union origins in Aquinas' concept
of a just wage. This sense of ethical duty could also have been
taken advantage of in instituting the new discipline of
dominance.

My suggestion that foremen could deftly utilize miners'
emotions to control their work behavior may at first seem un-
likely—where would the managers learn such skillfulness?
Ironically, all foremen came from the ranks of experienced
miners. The mine owners were actually forced by state regu-
lation to hire a man with "foreman's papers." Foreman's papers were earned by passing a series of tests; e.g. the fireboss
exam was the first step. The other stipulation was that the
candidate for foreman have at least five years of underground experience. In a fashion undoubtedly unexpected
by the owners, and probably not even recognized by them,
this practice of hiring former miners as foremen appears to
have ensured that optimal work was obtained from the mineworkers. This was particularly true during and following
the transition from hand-loading to machine-loading when
supervision became more intense.

Since foremen were former miners, and likely sons of miners, they shared work values with the men they supervised.
This included understandings of how to act and react during
miner-boss interactions—what emotional practices to follow.
Using Bourdieu's notion of *habitus* (1977, 1984), I would
point out that both were likely to have grown up within the
same *habitus*, which included not only work values but also
the emotional practices by which someone in authority might
make acceptable demands on those under him. They were
individuals endowed with, and interlinked by, the same emotional and value-laden dispositions. So managerial manipula-

tion of miners' emotions, such as the eagerness to please, fear of deprivation, pride, and shame, could occur through a cultural understanding of such emotional practices and would likely be so natural as to be unconscious and unarticulated.

From my understanding of the men who made the changes in worklife from the loosely supervised hand-loading to the closely supervised machine-loading, they did not rebel. Their work values allowed them to see opportunities in the new methods. More importantly, the emotional practices involved in work made them vulnerable to the exploitative practices of authority. The foremen knew in a non-calculating sense how to manipulate the men's emotions by having been exposed to the same *habitus* themselves.

6

Work Is What You Make It

... sometimes they'd have us run double shifts. If [the fore-
man] asked, we would help them out. Sometimes I'd come in
overtime. They'd give me a holler, and I'd go give 'em a hand.
—*former handloader*

I have been haunted by troublesome thoughts since writing
the previous chapter. I've regretted several processes in
which I felt bound to engage in order to transform my field
experience into a written analysis. Those are the processes of
objectification and of imposed perspective.

I felt that the process of transforming the vocal messages
of my acquaintances and friends into data for analysis was a
distorting act. Like Portelli (1984:115) I saw this as a "freezing
of the fluidity of words as an arbitrary point." Thus, something that was dynamic and spontaneous—perhaps worked
and reworked with many tellings —was made static and dis-
sociated from the teller. And like Portelli, I came to see it as
an inevitable consequence of my project.

There was a second process involved in this transformation to the written text that also disturbed me. As conversations and bits of conversation became data, they were freed
from the context of time, mood, dialogue, and even person.
These quotations were manipulable. I could stack, file and
columnize them in such a fashion as to make analysis
possible.

The struggle with issues of data manipulation and analysis
and of objectification have been resolved in different ways in
the past. Frisch (1989) suggests that the processes of inter-
vening between speaker and reader to rearrange and select
segments of informant narration is a necessary and honest

endeavor, like that of a craftsman who has a "feel for what is quality." Immersion in the material secures awareness and honesty. Dwyer (1982:277), on the other hand, opposes such a process. He considers it to be "hiding operations that they have performed on the text" that "blind the reader to fieldwork confrontation itself and thus deprive him or her of the chance to seriously criticize the author's work." Mishler (1986) and Crapanzano (1985) also express the opinion that the interview exchange is of paramount importance.

Oring (1987:260) is less concerned with presenting the ethnographic confrontation. He appears to view the life history as a synthesized monologue. Therefore, he considers analytic efforts to reduce them to data to "vitiate the generic motivation of creating a voice allowed to speak without interruption." He also expresses the concern that "an analyst is at risk of commenting less significantly about a text that already appears coherent, compelling, authoritative, and full of significance."

I found Dorothy Smith (1987:93) to be similarly concerned. She has cautioned social scientists in their priviledged position as writers to take care to not ignore the native's experience of the world by constructing "a sociological version which is then imposed upon them as their reality." We must not extract from their conceptual framework what fits with ours.

Faced with these choices, I had originally considered the method of Oring as exemplified by Wallace Black Elk and William S. Lyon (1991). I felt that the voices of informants could be presented as monologues, with my dialogic interferences presented in a reflexive analysis.

Ironically, I did to a large extent utilize a conceptual framework from which I extracted and analyzed their conversations. This involved the other process that rasped my conscience—the imposition of my perspective on the data in order to produce a meaningful analysis and interpretation. Although I included as many voices as feasible in the writing of my text, I still used my authority to arrange them, select them, and interpret them: a monophonic authority thinly disguised. Smith (1987:93), Clifford (1986:15), and Rabinow (1986:243-47) have cautioned against this practice, but when power and exploitation of a group are the issues being analyzed can an exception be made? I have felt so; however, I have tried, as a part of this chapter, to present as an alternative the natives' point of view with less secondary interpretation.

When I imposed my interpretation on the informant responses I was blind to the fact that I had inadvertently suggested that I was more astute than the handloaders in being able to detect the imposition of force. I wanted to point it out to them as well as to my academic peers. For any who were exposed to years of being improperly referred to as "dumb miners," this was likely taken as an insult. I want to assure those natives of the region of study that this was not my intent. Power and exploitation are complex phenomena that are not always easy to discern or to understand. That is why I feel that the misuse of power to exploit is a phenomenon that needs to be reported wherever detected. Although I detected an abuse of power here, I realize that the analysis of power has produced many theories and explanations among which mine is but one.

After writing this analysis, I decided to seek critical appraisal from informants. However, I felt anxious about doing so. I sensed that the analysis would appear alien and perhaps threatening to them. I side-stepped my anxiety long enough to present a draft to an informant whom I felt certain would give it fair critical appraisal. I sensed disapproval when the paper showed up two weeks later inside my door with a one-sentence note. I arranged to talk with him about the paper about a week later. Early in our conversation he told me: "Some of those handloaders, I'm sure, would resent it, if what you said were brought up to them. They were taught to give a full day's work." He then went on to describe in some detail how the desire to work hard was internal and "a matter of fact." As I listened again to his celebration of the work ethic, of his enjoyment of working hard, I came to realize that perhaps his perspective had as much validity as mine. This was a conclusion that had also been drawn by Light and Kleiber (1981:175) concerning the perspective of their informants during their research in a women's health cooperative: "By initially denying the intimate relationship between researcher and researched, we prevented ourselves from learning from the perception of collective members. We set ourselves up as experts and denied that collective members had crucial contributions to make in the observation and analysis of their own organization. Their perceptions as insiders were different from ours as outsiders, but they were no less valid."

In what follows I have sought to understand and to explain the insider's view of work. I reviewed all of my relevant interview notes following recent conversations with two informants. One informant was the former mineworker and mine boss who critically evaluated the paper. The other was a retired garment worker who in an interview months earlier had made the statement: "I think that work is what you make it." This sentence spoke of personal power, and I felt that to understand it might provide the key to understanding satisfaction in labor. I would first like to introduce this analysis with some mention of the probable work attitude of European immigrants to Pennsylvania.

Most of my informants were first-generation descendents of European immigrants. During the late nineteenth and early twentieth centuries these immigrants made their way to America in search of work. Bodnar (1982, 1985) found that those who stayed had a set of work values based heavily on a fundamental sense of realism. Theirs was a life preoccupied by survival strategies and family welfare. The world was an unstable place where the welfare of the family was tenuous— job security, job safety, and decent wages could not be taken for granted.

Bodnar also found that these immigrant workers socialized their children to accept, and even to pursue, steady, industrial jobs rather than to criticize or challenge the industrial system. My talks with these children, now themselves retired, would seem to support that assessment. In accepting the idea that an industrial job was not only an acceptable condition for living, but a desirable one, members of the last hand-loading generation were able to fashion meaning out of a structured system that outsiders look at as being imposed upon them.

I spoke twice with the retired garment worker in trying to better understand what she meant about work being what a person made it. The statement sounded as if she suggested that the worker had the power to choose between options in the workplace. Her opening remarks indicated to me that I was right: "What I meant was that you can either make it a horrible or a good thing: You could be cooperative and make it a good thing or buck the whole system and make it a bad thing. Some are never satisfied no matter what, whereas I used to take things with a grain of salt." I asked her to define what she meant by a "good thing": "I meant that the company would meet you half-way, where if you kept buckin' the system they made it miserable. For example, if they wanted you to work overtime—I never liked to work overtime, but I would anyhow. If you did, and later if you wanted some time off, you were more likely to get it than someone who refused overtime." Through the example she showed that being em-pathetic to company needs gained her flexibility in schedul-ing later. This approach is both empathetic *and* pragmatic. Certain aspects of the work situation are taken as given. To make the mistake of continually fighting to change them is to make your life miserable. The former mineworker and boss who had critiqued my paper had similar sentiments:

After a few years working in the mines I learned to respect the [older] men. They were genuine and hard-working. If they told you they were going to do something, they did it. The old men would come in [overtime] for you. They had that old pride and caring— they knew that the [company] light bill doesn't get paid automati-cally. Of course, they wanted the money, too.

There was very little bitchin' and moanin' about the company—and there was no boss around so they could have. It was just that nobody needed to gripe.

Older guys might not a liked it [working hard], but I don't think they even thought about it. There was no thought about machinery—it was, "Where's a good pick?"

According to these insiders, force and exploitation were never issues. It seems that they attained a level of satisfaction in laboring that they valued. They were pragmatic in two senses. They saw the company as providing the necessary means to a livelihood—to living a comfortable life. At the same time, they chose to create a system of meaning—a source of satisfaction—within a structure provided by the company. There were pride, dignity, rules, traditions, and sanctions—all within a work situation framed by industry.

There was pride, not only in physical ability, in providing for one's family, and in being a good worker, but in workmanship itself. The former mineworker told me: "The older guys had pride in workmanship—setting a prop [timber to hold the roof up and to warn of possible cave-in] right and knowin' that it would hold the roof. They'd know where to pick the coal, pick it with the grain. They'd know how to sharpen the pick. It was like it was a craft—like making this tabletop, getting it smooth under your touch." With the introduction of machine-loading, there was pride in crew production:

I remember we had a competition in runnin' shuttle cars. We'd look for the chalk marks [on the frame of the shuttle car], at how far the shift before us got. Most of the time it was a pride that drove us . . . It wasn't a force from the boss—it came from the men themselves. I remember when I was a mineworker, the miner operator [continuous mining machine operator] was an older man. He'd yell at the buggy runner [shuttle car driver] for bein' so slow and for not getting off the shuttle car to help at the face to speed things up or to just help the crew.

There was dignity in providing just labor for an agreed upon wage. There was also a sense of dignity in discipline, skill and endurance. My informant used the analogy of football: "It's like the Penn State football team: Joe Paterno believes in hitting hard—no razzle dazzle. You might lose 42 to 20, but you'd still have pride." The rules included loyalty to the company, providing a just labor for an agreed upon wage, earning as much as one could, and providing as much as possible for one's family.

The traditions included being taken into the mines, apprenticeship, and indoctrination. I recalled a statement my informant had made to me months earlier. It also spoke of a tradition: "I'd be there in the wash house soapin' up, and here'd be all these guys old enough to be my father or grandfather with fingers and tips of fingers missin', with their beer guts hangin' out. They'd growl: 'So, kid, you like to work in the fuckin' mine, eh?' Years later I found myself in the wash house growlin' to the younger guys: 'So, kid, you like to work in the fuckin' mine, eh?' That's when you know you made it."

The sanctions included knocking. I heard more stories of mineworkers knocking other mineworkers than stories of bosses knocking mineworkers. My informant told me: "You would embarrass them to help you because you were low rate." There was also the action of "dumping a buddy." My informant described it as it occurred in a mechanized mine: "If you had a buddy who wasn't good, you tried to dump him. For example, if you were running shuttle cars with someone who was slow." [Two shuttle cars serviced one continuous miner. They alternated trips from the coal face to belts that took the coal from the mine. Roof bolters also worked in pairs and the continuous miner operator had a helper]. A former miner gave me an example from handloading days:

The boss would give me an old man as a buddy. Those old men— they were worn out. All their sap was drained from years of hard work. We'd load a car, and the old man would want to stop for a drink of water. Then we'd work a little more, and he'd want to break for a pipe. I told him, "The trip's goin' to be comin', and we're goin' to lose a car." They would only drop you off as many empties as you had loaded cars. Finally, I told him to get out, and I worked by myself. I took the cars like I had a partner workin' with me.

There may have even been certain axioms involved in the work attitude. For example, I saw a sign above my informant's desk in his den. It read: "Good things come to he who waiteth as long as he who waiteth worketh like hell while he waiteth."

As in previous discussions, my informant took time to speak about the attitude of young miners who entered the mines in the late 1960s or 1970s. He described how they lacked the desire to work hard. If equipment broke down they often just sat around. It sounded to me as if work had lost meaning for them—as if work had become a painful process to endure. Without an appreciation for the traditions and satisfactions of the old work attitude, their work life became aimless and meaningless. My informant pitied them:

I still say that a lotta young guys were cheated. They thought that was the way it was supposed to be. They never learned what caused them to get to this time. They didn't know about hard times, hard work, the depression. All they knew was: "I can walk off the street and get one hundred dollars a day, and the union backs me. I don't have to work all that hard." Today, if you can get *out* of doin' something—that's smart! [something to be proud of].

I recalled his saying in earlier interviews:

They changed the rules so that you didn't need to have someone sign you in for a year as an apprentice. The company could hire as many as it wanted . . . These newer men just didn't understand. These guys were brought in when times were good—it was big bucks. Who knows, maybe some of them would have accepted the old ways if they would have had the chance to learn. In the old days you were indoctrinated to work at everything when you were in the mines. The newer fellas were usually indoctrinated to one job— that's all. That would be their job description. It wasn't their fault. They came into the mines when times were good and never got the opportunity to gain from the experiences of the older miner.

He also told me: "The work attitude was a handed down thing—a real individual need to do good."

Handloading under a low roof, in a wet section, or where coal had a high level of impurities made the job deplorable. Courtesy of the U.S. Department of the Interior.

In this final section I will take a closer look at the nature of hand-loading and of machine-loading. Had Goodrich, and have I, made assumptions about the nature of each that exaggerate the importance of the transition between them?

The frustrations of hand-loading are undeniable. Miners had to deal with the vagaries of nature, with the shortcomings of car allocation, with the lack of pay for deadwork [work such as cleaning up rockfalls, bailing water and laying track], and with the fussiness of removing impurities. I will review in more detail the nature of these categories of frustration.

Concerning the problems with nature, the miner had to tolerate such possible frustrations as standing water, narrow seams, thick impurity seams, and loose roof. As Dennis, Henriques, and Slaughter (1969:65) point out, this was not like factory piecework: "Piece-work of this kind is paralleled to some extent in other industries, but the direct confrontation of the miner with nature brings an element of unpredictability which is unique."

A miner's production was limited by the number of cars he was allocated, and production decisions were often limited by the time available between car pickups and deliveries. Miners might be rushed to get their cars filled on time or might be idle without choice if car turnaround was slow. Likewise, the cutting machine schedule could dictate the pace of work.

The miners' contract included the right of companies to penalize the miner for dirty coal. Most often this was enforced by dumping the cars of dirty coal onto the waste dump. Another tactic was requiring the miner to clean the suspect carload of coal when he came from the mine. Repeated offenses could lead to days off without pay and eventually to dismissal. Most mines had coal inspectors outside who went through the loaded cars. Some mines had coal inspectors inside to check the coal as men loaded it. Among the impurities to be removed were boney [a carbon-containing shale material that resembles coal in appearance], some types of high-sulfur coal that lay in a seam above a boney seam, pyrites, and in some mines "bugdust" [the fine coal from the cutting machine].

Just as the hand-loading of coal was probably less satisfying than a cottage industry, the machine-loading of coal was probably less monotonous and restricting than a factory job. During my trip inside Rushton Mine in the fall of 1988 I watched a production crew at work in a forty-two-inch coal seam. I was struck by how much the activity resembled that of excavation work. These people operated heavy equipment and drove truck-like shuttle cars. The workers took pride in their skills with the heavy equipment. Workers had time to talk with each other between the times when their operations were necessary.

The elements still had to be contended with, but two features were different from hand-loading days: high capacity pumps, roof bolting machines, and mechanized scoops were often utilized, and the crew took on the challenge to beat the elements without worrying about the day's pay. In fact, two former mineworkers independently described their jobs to me in terms of an army battling the elements. These underground excavation teams represent to me a less regimented group than either an army or an assembly line. This does not mean that monotonous jobs don't exist in modern mines, however. Workers had described to me the frustrating jobs of belt-cleaning and sump-cleaning.

Being concerned, still, about the intensity of supervision, I asked a former mineworker and crew foreman for details about the responsibilities of the face boss [production crew foreman]. I learned that the "face boss" had responsibilities that took him away from the crew for hours every day: "It depends on the mine. They may have duties to inspect haulage-ways, belts, and so on. Then they have to 'pre-shift' [inspect their section of the mine in preparation for the next shift] and 'on-shift' [inspect] for their own crew." These details of hand-loading and machine-loading call into question even further the importance of "freedom" to the miners.

It also calls into question my analysis. While I detected little sense of loss of freedom by the handloaders, I suggested that freedom was nevertheless taken away. If the conditions of handloading were barely tolerable and if machine-loading was other than factory-like, then perhaps the discussion of coercion is ill-fitted to this topic.

I hope that this postscript has decreased the level of my objectification and my imposition of my perspective on this subject. The result is a less conclusive and less neat analysis, but one truer to the thoughts of the residents and to the complexity of the subject matter.

7
Why Worry?

Water is dangerous in the mine; gas is dangerous in the mine; and cables are a danger in the mine.
—*retired mineworker and former member of the mine rescue team*

It is necessary to introduce the nature of mine dangers before attempting to describe the natives' emotions involved in living with those dangers. I considered using mine safety literature or an interview with an administratior in the MSHA office; however, I eventually rejected these alternatives so that I might bring more primary sources into the analysis: testimonies from a wide range of mineworkers, my own personal experiences, and archival newspaper accounts. Although this information may be less detailed or complete than other sources, it is more genuine because it was experienced. Some dangers have been prevalent throughout the lifetimes of the retired miners while others changed as the mines became more mechanized.

The explosions leading to disastrous loss of life are the phenomena that people think of first when imagining coal mine danger. The central Pennsylvania bituminous coal region has had twenty-two disasters in which five or more men were killed. These include: the Rolling Mill Mine (112 died in 1902), Reilly No. 1 (77 in 1922), Clymer No. 1, also called Sample Run (44 in 1926), and Sonman E (63 in 1940).

Methane is the chief culprit in mine explosions. It is released from the coal as the coal face is freshly exposed. If it comprises 5 to 15 percent of the air it is explosive. One of my informants told me he used to demonstrate this to other miners by taking a chunk of freshly mined coal from a car outside the mine, breaking the lump, and igniting the methane that was released. According to informants it could be ignited by many sources: trolley line sparks, electric motor sparks, carbide lamps, sandrock striking against rock or steel, the friction caused by rapidly moving gas against mine walls, and sparks from cutter bits hitting rock. Former miners told me:

When I was a fifteen-year-old kid I caused an explosion. I tried finishing a cuttin' job with dull bits. The sparks from the cutter caught the bug dust [and gas] on fire. The fire danced all around the room. All the hair that stuck out from under my cap got singed. It must of made fire out the air shaft 'cause the boss came runnin' to see if I was all right.

There was a gas explosion. The arcwall machine was trammin'. It would arc along that hot wire and explode each time it hit a gas pocket. All that black dust started to look like red pepper, and I couldn't breathe. I laid down and got some air near the ground.

A survivor of the Reilly disaster described the way his open-flame carbide lamp was behaving the day of the methane and dust explosion: "Since the mines were only working two or three days a week, gas would build up in there and settle in certain spots. Sometimes gas at the face would cause the flame on my carbide lamp to flicker." A newspaper account (Barnesboro *Star* 1921) describes what happened to two miners who ignited a pocket of methane: "An explosion of gas occurred in the Watkins Mine No. 3. The dead are George Olichlick of Bakerton aged 42 and Nicholas Antonnuccio of Bakerton. Both bodies were charred almost beyond recognition. Lamps ignited pockets of gas."

The level of gas in coal varied—some mines are said to be "gassy" or gaseous. Certain conditions enhance the likelihood of methane accumulation: mines sitting idle for three to five days a week (as was the case for many mines in the 1920s and 1930s) and the high productivity of the era of machine mining. The more coal chewed from the face, the more methane released.

Methane could be detected by using a "bug light," a flame lantern that would flare more brightly in the presence of methane. Handloaders trusted "fire bosses" to check methane levels during their rounds. In modern mines each mineworker carries a detector, but not everyone checks them regularly. More recently, mining machines are required to have a built-in methane detector that will shut down the machine when levels get high.

Another gas danger involved mining into sections of old gas wells: "There are a lot of old, abandoned wells around that aren't on any maps. Bethlehem No. 33's long wall hit an old gas pipe and sheared it right off. The gas came out of there with such force that they had to shut down operations for three days."

The danger from an explosion went beyond being burned in the flash fire. The oxygen would be rapidly converted to carbon monoxide and carbon dioxide. Carbon monoxide is, of course, a poison, but both gasses tie up oxygen in a form unusable by the human body. Mineworkers now each carry a self-rescuer—a canister worn on the belt that contains a breathing apparatus that can provide oxygen for a short time. One man told me that if the carbon monoxide level is high the old-style self-rescuer (which converts carbon monoxide to oxygen) can get so hot that it can burn the user's lips.

In addition to the large scale methane explosions, hand-loading miners had to deal daily with explosives in shooting down the coal for loading. Holes were drilled into the coal seam and either black powder or dynamite was placed in the hole. For black powder, powdered coal was packed around a long needle that reached into the tamped explosive. The needle was carefully withdrawn, and a squib was inserted and lit. A squib is a piece of twisted paper—about six inches long and roughly one quarter the diameter of a pencil, the last third of which is impregnated with black powder. The squib smoldered after being lit, the hot coals of the burning paper moving slowly towards the explosive tip. Once the tip was reached the squib shot like a rocket into the powder, igniting it.

When it worked it was a fine system for bringing down the coal. Unfortunately, the squibs did not always fire within the expected time frame. Then the miner was faced with the decision of whether to set new charges or leave the mine for the

A miner drilling a hole for an explosive charge. Misfiring squibs
led to many accidents. Courtesy of the U.S. Department of the
Interior.

day: "When I first started, they had black powder and squibs.
If a squib didn't go off, the law was you had to leave that
place until the next day. I seen where the squib just lay in
there smoldering and went off the next morning." However,
this meant that the miner did not get paid for lost time since
he was being paid by the ton for the coal he loaded. Most, if
not all, the former handloaders to whom I spoke had replaced squibs that did not fire. I also heard several stories of
men losing their eyesight and hands from checking squibs:

A Polish guy who lived near the powerhouse around 1924 was killed
in No. 9 when the squib didn't go off. He went up to it with a carbide lamp. When the black powder blew, it blew the coal at him. It
was the coal that hit him.

We weren't supposed to remove a charge if it didn't blow, but we did
it anyway.

[He] was shootin' with black powder at Brawley Mine. He tamped a hole around a needle and put a squib in. The powder hadn't gone off, so he went to check—that's when the squib ignited. He was a good old man.

My Lithuanian neighbor lost both of his eyes checkin' a squib that didn't fire.

And from the newspaper (Barnesboro *Star* 1921) came this description of an accident victim at Logan No. 5 Mine in Carrolltown: "The unfortunate man was badly blown up. Both eyes were blown out of his head, and several fingers and flesh from his body were missing." The same economic incentive to prevent lost time led to the practice of using short fuses when dynamite was used to shoot down the coal: "Them old miners used to use short fuses. You don't want to wait—you ain't makin' no money." Related to the dangers of shooting down coal was the practice of shooting coal from the solid. In this procedure, the step of undercutting the coal was skipped. Holes were drilled into the solid seam of coal, and the explosive was ignited. This dangerous practice was not common. I was told of one miner who carried it out to both save time and to break up boney layers into indistinguishably small pieces so that he could get away with loading it along with the coal. A former handloader told me why he avoided this procedure: "The coal would shoot out at you and up towards the roof where it would put pressure on the roof."

Dix (1977) demonstrated that roof falls caused more deaths and injuries than mine explosions for West Virginia. In talking with people from the central Pennsylvania coal region I found that roof falls (also referred to as rock falls) was a topic most widely discussed in terms of mine danger. People spoke of losing a loved one or a buddy to the roof's collapse. There are many accounts of near misses while mining:

Once we were setting props, and my buddy went to get a spacer to fit on top of the prop. When he come back, the roof was sittin' on top of the prop. [When we saw that] we started scrambling, and that roof started fallin'. And it kept following us as we ran—all the way back.

Once when I was working with my brother at Colver, a small chunk of roof fell near me. I tapped the roof with my pick and heard a dead thud. I yelled to my brother to get out. He leaped and fell on top of me, knocking me over. The whole section of roof fell with a deafening roar. We were in such shock that we couldn't work. We asked the foreman to go home, and he understood and let us go.

We cut just under the boney [layer of rock impurity]. When we got back in here where we had water layin' it would push the boney down. One time the rock and boney was fallin', and me and a fellah who worked with me run. And it was hittin' our heels, you might as well say, as we was runnin' out.

One time . . . they wanted my brother to do some work by the conveyer. I told my brother to work somewhere else because the roof was bad there. I tried to tell the foreman, but he said the work was O.K. My brother worked somewhere else. That roof came down with a boom. It took a month till they cleaned that place up. It might have killed my brother.

There were also many stories of broken pelvises, skulls, and backs:

I worked inside the mines from 1920 until 1941 when a rock fell on me in 37 Mine. It was a piece of rock from here to the door [about ten feet] and from here to that wall [about twenty feet] and about two and a half feet thick. I was caught between that and an iron rail—smashed my hip. I was in the hospital for one and a half years and didn't try workin' again until 1944 when I started at Wilmore.

My brother got his back broke in '26. He was never much good after that. His back bothered him. He wore a cast for a year, year and a half. A spar off a clay vein fell on him—four or five vertebrae got broke.

The way the accident happened, we were just going to set our props, and the first thing—bang!—just like that. If my buddy had just waited one minute. It smashed him right up. It was a roof fall. My buddy was killed. He was only twenty-five years old. I had my fourth and fifth lumbar fractured. It put pressure on my sciatic nerve. They thought that I'd never get out of a wheel chair. I had to learn to walk again—was out of work for four years.

I was in an accident in 1939 and was off for four years. My pelvis was crushed. I can see that rock coming down yet. My hard hat got knocked to the side of my head. It's a good thing because that hat was smashed flatter than a pancake.

[My dad] got killed while demonstrating the [Jeffrey arcwall cutter]. A thirty-eight-pound piece of rock hit him on the head and killed him instantly. It crushed his skull and his brains came out.

In those days [depression] they was short on timber. There were so many men with broken necks, broken backs, and broken skulls for the simple reason of rockfalls from not havin' enough timber.

Roofs were unpredictable. Some men had adages about safety under certain roofs, but other men could point out to me situations where these adages didn't hold. Slate roof, which is normally safe, could come down in one huge chunk, collapsing mine posts under it: "A slate roof can be dangerous, too. It can fall without warning—and in huge pieces, too . . . I seen a fourteen-inch thick layer of slate, twenty-five feet long and a piece six to eight feet wide fall just like this [dropped his arm on the table] without any warning. When rock falls like that, and it doesn't break, it can knock the timbers out—they'll slide out." Likewise, sandrock could fall unexpectedly even though it was generally known to give much warning with loud cracking sounds: "Oh, it'll crack, and long before it falls. That sandrock in Barnes's 20 surprised a lotta men, though. A couple had broken backs from bein' hit." Men told me of seeing or experiencing falls of slate, shale, sandstone, boney, soapstone, and concretions in sedimentary layers that could fall out of an otherwise supported roof of sedimentary rock.

Men argued vociferously on the merits of roof bolts (also called pins) and timber supports (also called posts). Posts represented a traditional means of roof support and involved placing a beam vertically under a spot on the roof. Timbering was done by individual miners during handloading days. It involved cutting the post to the appropriate length, setting it straight and pounding a "cap-piece," a broad wooden wedge, between the top of the post and the roof. In the low coal of the central Pennsylvania region this was most often done from a kneeling position. During the 1950s and 1960s a timberman set posts for the mining of coal with machines like the Clarkson loader, the Joy loader and the Jeffrey Col Mol. As each new face of coal was exposed, timbermen would move in and set posts before the machine was brought to the face again.

By the late 1960s, roof bolts were used to provide a more convenient form of protection for use with machines. A roof bolting machine drilled holes up and into the sedimentary layers, and then bolts were screwed into these holes. The theory is that the bolts strengthened the roof by anchoring it into upper strata. More recently "resin bolts" began to be used. These used a resin placed in the hole prior to the screwing in of bolts. The theory was that the roof was strengthened by cementing sedimentary layers together, analogous to creating plywood from many layers of thin material. Bolts ensured unobstructed passageways for the shuttle cars carrying coal and for the continuous mining machines. Many older miners, however, saw danger in abandoning posts. For example, one retired miner told me of seeing rock falls in which pins would be seen sticking up out of the rubble: "You'd see roof falls where the pins were stickin' up out of the [rubble]. Where I worked, the mines had soapstone—it would reach up twenty to thirty feet like this [convex dome]. Those five-foot bolts only reached up five feet." Here water had seeped around a section of roof to form a dome. This dome could slip out, and the bolts often would not be long enough to anchor it into higher rock.

Horizontal metal beams also have been used to support roof—generally held up with posts. These had their own shortcomings, expense being the major factor. They were valuable enough to warrant recovery at the jeopardizing of a miner's safety in the process:

They paid men to get the iron cross bars in the headings. I seen places where we had ninety-pound iron [ninety pounds per foot] where it would be bent like this [moved his hands in a crescent].

They used to use steel railroad beams as cross supports, but they were too dangerous. When weight came down on them they'd snap. Sometimes they'd turn crooked, and they could crack anytime and kick out. If one should hit ya it would kill ya dead.

Removing coal pillars usually involved the collapse of roof. Pillars were the blocks of coal remaining after the coal was mined in a cross-hatch pattern. The weight of the roof on these pillars often compressed the coal, making it easier to remove. Many men preferred mining the "squeeze coal" since they were paid more per ton by the company. The probable collapse of roof made such work more hazardous: "Sometimes the same men that mined the rooms would pull the pillars, but sometimes the better men were put on pillars. It was more dangerous —you had to be more precautionary about roof conditions. There were more accidents than with advance work." In modern mines the removal of pillars is called "retreat mining" in contrast to cutting the tunnels, which is referred to as "advance mining." The size of the roof fall is usually greater since roof bolts hold up larger sections of roof. Examples of the danger of modern roof falls is evident in the following accounts:

When it came it looked like a big curtain droppin' down. Eight sixty-foot entries all dropped in less than one minute, and there was a machine underneath it.

The foreman told me to remove a pillar. I knew the roof was bad there so I said, "Why don't we go down where we can load twice as much coal? There's only two hours left, anyhow." He said: "No, my supervisor said that they want the pillar out of here." So I started running the Lee Norris [mining machine]. As I worked I started to feel some flakes falling from the roof. I left the machine and got out of there. The foreman went after the machine, and the roof fell on him and the machine.

In 20 Mine when we was pullin' pillars, areas would open up twice the size of a football field. The mine foreman was deathly afraid of cavin'. He saw that we were about done, and he headed out. He was way down the heading and just opened a wooden door used to block air when the roof caved. The pressure traveled all the way down there and took off that door.

When the roof really starts makin' noise we stop and watch it. One time when I was drivin' buggy, I got back into the pillar area just as it went. It took the canvas right off, and it took coal outta the shuttle car and dumped it down my collar.

When that roof goes, it sounds like steady thunder for a while. When you're takin' out a pillar sixty feet on a side, you're still down in there pretty far. First you go down the middle and then you start cuttin' off the corners. The foreman we had insisted on gettin' out the last bit of coal. Gettin' that last buggy-full was really scary.

I remember the first time I was in a pillar fall. We collapsed twelve or fourteen rooms. There was nothin' but posts as far as your light would shine. I wasn't smart enough yet to be too edgy. There was a gush of air, and one of the canvases blew over my face. It knocked me over. It scares the shit outta ya!

In those cases where men can survive being buried under the rock fall, they become vulnerable to exposure as they lie in cold conditions.

Some miners avoided work in high coal [six feet or higher] because of roof fall conditions. The rocks would come down with more force and from the ribs as well as the roof. Also, the rocks might bounce and roll for greater distances.

Boney, a shaley impurity in or next to the coal seam, was described by several men as the hardest roof to work under. Many companies tried undercutting boney layers rather than mining the boney and coal and later separating the boney out. This roof was very fragile, and it did not adhere well to upper layers because of a loose layer of "top coal" that lay between the boney and the upper sedimentary rock. The following account indicates the problem it presented:

The boney [thickness] ran from eighteen inches on up. Then there was a little bit of top coal. That would loosen itself from the boney . . . We had pins and props and everything else put up, but it would break loose and come down. Once when I was boss we was takin' out stumps [blocks of coal supporting the roof], and it started cavin'. I shouted, "Let's get goin'," and we run out. The assistant [mine] boss come by and said "I want that stump out!" We went back, and all that boney come down and hit me and [my men]. It felt like you was smothered. Boy, I couldn't get over that—that assistant! [One of my men] got covered up and got scared. He never went in the mines again. We coulda got hurt bad—we coulda got killed. It's a good thing that it was boney and not solid rock.

The underground transportation of coal, supplies, and men had long been a major source of danger in the mines. Brophy (1955) gave a description of an accidental death in a central Pennsylvania mine near the turn of the century: "His hands slipped, and he wasn't able to get a handhold, and he fell down in front of this car and was killed almost instantly." Cars were first moved by mule, but later electric locomotives (called "motors") were used. The dangers with these more powerful motors included collisions with humans and with other "trips" (trains), derailments leading to men being "pinned" against a wall or another car, runaways on hills, and fingers caught in mechanisms like couplers and wheel spokes:

I was spragger [brakeman] one day when the [ventilation] door flew shut, and the motorman hit it. Another time some cars run away and hit the machine next to me and pinned me against the rib. I just got bruises.

A "snapper" was the same thing as a spragger. He would "snap" a wooden rod in the spokes of a wheel to brake the car. There was a lotta busted knuckles.

Another man jumped off a motor and ran up to throw a switch to prevent the trip from headin' onto the siding. He fell forward as he throwed the switch and the motor run o'er him and cut his body in half.

There was a guy tryin' to beat his time to leave early. He was drivin' out a small motor. He knew there was a big locomotive comin' in haulin' cars. He thought that he could make it to the switch and pull off. He met it on the curve and got busted all to hell.

Supplies and a few men at a time were moved by "jeeps," single trolley cars powered by their own electric motors. I experienced two realistic fears as I rode in a jeep at approximately thirty miles an hour in Rushton Mine. I worried about my head hitting the low hanging and variable height ceiling and the threat of suddenly running into a stationary car or motor on the tracks. I later heard unsolicited tales of these very occurrences:

I never got hurt bad enough to get compensation, but once I got hurt driving motor. It was so dusty that I couldn't see ahead of me, and suddenly I seen something shiney ahead. I hit the brakes but couldn't stop before I hit some cars parked there ahead of me. Luckily, I stood up because a piece of the metal pipe I was carrying came flying forward and hit me in the meat part of my leg. I got a heck of a bruise. If I had still been sittin', that pipe would have hit me in the ribs and probably would have punctured [my lungs]. Who knows—I might have been dead!

I tried catchin' up with somebody in the mine. I was drivin' a jeep as fast as it would go when my head hit a low-hangin' part of the roof. It pushed my neck down into my back with a jolt. It was my own stupid fault. I have two chipped vertebrae and a pinched nerve. The doctor said that if they operate there's a seventy percent chance that I'll be paralyzed in my left leg.

We moved the cutter on rails, and they had the rails jacked up. I told 'em I wasn't cuttin' until they moved the track back down where it belongs. There was a dip that the tracks went through, but I wasn't worried about it. The tracks were covered with water, but I [had] walked in that way, and the clearance to the roof stayed the same. I started to ride the cuttin' machine out, and it caught me against the roof and broke my back. Here, they left two ties under the track in the water hole. I made a dive to get out but it pinched me. There was only nine inches of clearance over the machine. They didn't know how I got out. I was lucky—very, very lucky. I walked right through that water hole, but I didn't know that the tracks were jacked up.

Coal cars could also be moved by means of a cable [called "rope haulage"]. Here, the dangers included cable breaking, becoming pinned between cars and walls, and accidents in riding out on the cable. In the No. 10 mine of Peale, Peacock, and Kerr, two separate incidents of fatal pinning were remembered to have occurred in the same year: "My father was killed in '25 on the rope haulage. That was when they were hauling coal [underground] from No. 10 to the No. 9 tipple. Six months later our neighbor, Mr. Lang, was killed. That was on the long haulage also." In modern mines, coal is moved all the way through the mine and outside by means of belts. Mineworkers were once expected to ride the belts, a practice that led to head and limb injuries from hitting stationary objects. Arble (1974:42) has described such incidents: "Another man was riding the belt at high speed, five hundred feet a minute or better. It was illegal, but no one will scuttle like a crab for a half mile under forty-two-inch roof when he can hitch a ride on the belt. A small cave-in ahead allowed the belt to pass through, but not a rider. He slammed into it head first and was killed instantly."

When continuous mining machines are utilized, the coal is usually transported from the mining machine to the belt by means of an electric shuttle car [called a "buggy"]. During my trip inside Rushton Mine, a student, our guide, and I had to "run" between cross cuts [side tunnels] before the shuttle car came through again. The guide told us to follow him. It wasn't easy. We were under about a forty-four-inch roof. Roof bolts stuck down every four feet across the twenty-two-foot-wide heading. He suggested staying between rows of bolts. He sped along using his short pick handle as a cane. The young student kept pace fairly well. I soon found myself falling behind. My arthritic hip gave me some pain as I attempted to do a rapid side-step like that of a fiddler crab. I had to try to keep my knees bent, my back bent, and my head down. Even so, I sometimes scraped my back, and any time I turned my head enough to see where the other two were I hit my helmet on the roof . . . I took to grabbing my left leg with both hands to help lift it . . . We had the shuttle cars to contend with so we had to try to move faster so as to make it to a cross cut before a buggy came through.

The student and I were tired and hungry, so our bodies were less responsive than the guide's. As I tried to hurry, I found that I was less likely to stay in a straight line. At one point I lost my balance and fell, catching myself with one hand. The hand sank about four inches into black muck. Now I lagged even further behind. I got up and scurried more, panting at the exertion. It was at that time that I came to the low roof area. My feet almost tripped as the floor rose, and my back caught on the roof. I heard a rip in my jacket. My compatriots were out of sight, evidently in the safety of a cross cut. Meanwhile, the ubiquitous whining and rumbling of the shuttle cars grew louder. Despite repeated bumpings against the ceiling, I made my way out of the low roof area and to the cross cut. We sought the safety of two more cross cuts before we came to the trolley line. Later, I heard stories of men being crushed against a wall by "buggies":

I remember when [my brother] had his leg taken off in the mine. He was a boss and was tryin' to help some other fellas. Here a car came around the curve and pinned him. I remember sittin' in the hospital all night waiting—seein' what was goin' to happen.

I was standin' in a [cross] cut when a shuttle car came flyin' around. It pinned my leg against the rib [tunnel wall]. I was in the hospital for thirteen days. The hospital said it was massive hemorrhaging and a severe contusion.

With those buggies, there was so many killed and injured—pinned against the rib.

Some shuttle cars were powered by batteries. Others took power from a main terminal by means of a long cable. This cable could also be dangerous.

Transport equipment was not the only equipment that endangered the mineworkers with whom I spoke. There was danger in using all types of machinery in the mines. There were also mining machines with names like Joy loaders, Lee Norses, Wilcoxes, Clarkson loaders, Col Mols, arcwall cutters. Even more innocuous-seeming were water pumps, rock dusters, conveyor pans, and roof bolters. Close quarters and water made the likelihood of accidents even greater than would be expected in using such powerful equipment. The Wilcox miner provides a good example. A retired miner described its operation: "One of the first Wilcoxes out was bad. Those things'll jump back like a power saw." A former crew foreman later told me: "I remember bossin' guys at 20 [mine] with all their teeth on one side knocked out by old Wilcoxes. The newer Wilcoxes have hydraulic posts so the man doesn't get near the auger."

I heard accounts of accidents with other mining machines:

I went cutting on a Joy [loader] and was on it for about five years. It didn't have rails, but was mounted on a cat-track [caterpillar traction]. It was pretty tricky using it. My buddy on the track lost his leg from the knee down with it.

My buddy got hurt real bad one day. The mine operator said that he was goin' to move the machine and to get away from the tail. He started to raise the tail, and we heard a squeal. Here, this guy went and sat on the tail. He got squished up against the roof. He's still in a wheel chair.

I used to work with a man named Curly. After I left him—maybe a week later—he got caught between the [Col] Mol and the wall.

I still think I could have gotten [the continuous mining machine] out of there in time [before the roof fell on it], but my helper ran off. He left all those high voltage cords [power cables] which I might have run over.

The mining machines also created new problems with roof support: "When I first started we went into the face without roof bolts. We only had props along the continuous miner . . . One time we was workin' at the face under sixty feet of unsupported roof. The boss called us up to take blocks off the belt. While we were there we heard a loud boom. Where we woulda been, it all caved on the miner." The longwall could also give problems with roof support. The longwall was a five hundred-foot assemblage of a rotating cutting wheel used to shear coal from the face, a conveyor pan to catch and transport the coal, and a series of hydraulic jacks used to hold up the roof. The pans and jacks were moved forward as more coal was sheared from the face. One retired mineworker told me: "Towards the end of my time on the longwall the roof pressure started to break rock ahead of the wall [hydraulic jacks]. It was really hairy as hell." Before the advent of mining machines and longwalls, cutting machines were used to either undercut or overcut the coal seam prior to "shooting" it down. These machines presented their own problems. The blade, which was similar in design to today's chain saw, could cut into a leg. One retired miner told me of a worker running the cutting machine off the rails. It spun around and cut the leg of his buddy. Other dangerous machinery included pumps, rock dusters, and augers:

He was in there by himself [on a Sunday]. His jacket got caught [in the pump mechanism] and pulled him in. He got a pen knife and started cuttin' his skin. And he had to drag himself fifty yards to the motor. He drove himself out. We was playin' ball, and he shouted to come down. I remember all the blood—you could almost see his intestines.

Those old-time mine pumps had a big arm that swung out and back. If you weren't careful—ping! There goes your arm! I knew several men who lost an arm that way.

[He] worked as a pumper—got his pant leg caught in the wheel. After that he worked outdoors.

My partner got killed in 1967. He was using an auger machine in a three-foot-high coal seam. It pulled him right into the machine and crushed him all up.

Fingers were always prone to being caught in machinery or pinched between the machine and a wall or roof:

So many was hurt with conveyors when they got 'em. Lots lost fingers with the pans.

[He] had his fingers chopped off in a mine accident. He put his hand on the canopy of his continuous miner. When it went over a bump his fingers got caught between the canopy and the roof.

Water has always been one of the more insidious dangers in mining. It could lie quietly above the roof or on the other side of a wall—a potential flood. Men who ordinarily had no qualms about mining often had great fears of mining near "old workings" (abandoned mines) which would be filled with water:

Over at Westover two men hit into old workings and almost drownded. They were in a dip—they just got out. It woulda been a few inches from their noses. The guys laughed it off—they didn't realize how close to the end they was.

The only time I was scared was when we was gettin' close to the old workings of Sterling 6. Water was comin' up from the floor, and it smelled. I was runnin' the miner, and a stream of water about the size of your finger shot outta the coal. So we shut off our machines and got outta there. I was afraid the water would break through. They're supposed to leave three hundred feet as a barrier between the workings, but in the days when they mined Sterling 6 they used hand-drawn maps, and they mined differently. A miner might make a dog hole. That's a short tunnel maybe six, fifteen, or even fifty feet in—if he was gettin' good coal, even if he wasn't supposed to.

When No. 10 flooded that one time, the pumper was the only man in the mine. You could see his light way back there but there was flooding between him and us—deep water. He had to hold onto timbers and wires to get out.

[He] drowned in No. 9. He dug into old workings that had filled up with water. Somebody screwed up. When that happens, water fills up the room that you're working in.

Harvey was cuttin', and he hit old workings that had water in it. Harvey started runnin', and the water caught him in a dip. At No. 10 they didn't keep maps up to date. Now the law is that you have to drill four test holes with twenty-foot augers when you're near old workings—one on each side and two ahead. Down in Springfield 4 they was headin' toward the old Reilly workings. They didn't have good maps, and they didn't know where they was goin'. I was deathly afraid in there. You can't trust any of those old maps—like Reilly—to be right.

Mining near old workings was the equivalent of cutting into the side wall of an underground reservoir. An equally ominous situation is mining under a creek: "I helped the day that Peale's had a mine flood behind the Benedict Hotel . . . Inside, I had to go around through old workings to get to the place. The tracks were all twisted up. It drowned out pumps and everything. No. 9 never opened again. It didn't hurt No. 10 much. There was only about fifteen or twenty feet of cover where the [creek bed] cave-in was." The Johnstown *Tribune* on August 3, 1915, reported the following incident: "A mine catastrophe was narrowly averted last evening at Sample Run Mine near Onberg, about seven miles from Indiana. About 200 miners were entrapped when Sample Run overflowed its banks following a cloud burst and poured its muddy waters into the mine mouth . . . In water up to their necks, the men waded for three fourths of a mile while six ran back to an air shaft and reached safety. No one was hurt, and it will take several days to pump the water out of the mine." Water is also insidious because it can provide a superb grounding for the electricity used in the mine:

One time the creek broke through into the mine, and me and my buddy walked out with water up to our necks. We wasn't supposed to come out—there was an electrician there to tell all the men to stay back. They was afraid that the water would reach the power line, and the power wasn't shut off yet.

One boss told me to bolt a section of roof in water that was over the top of my boots. Do you know how high a miner's boots come? [he indicated with his hand that it was one half the way up his calf.] It was close to touchin' the level of the electrical box on the bolting machine. When you set the jacks to level it the water would touch it.

Once when we were drivin' a scoop [coal loading machine] up a twelve hundred-foot headin'. We shot the coal, but I had to wade up to my rear in water in a low spot to check out a short in a light. I got shocked. I was lucky I didn't get electrocuted.

The poor movement of underground air was an especially troublesome problem during the days of handloading. I've already mentioned the problem of methane accumulation. An additional problem was the fumes from black powder or dynamite, which were slow to leave after the coal was shot down:

It was about four feet high, but there was no air. The black powder'd kill ya. They weren't strict about inspectors in them days.

In the early '40s you didn't have a third the ventilation of today. You fanned the smoke out with a shovel after you shot.

The old-timers worked in smoke—there was no ventilation. Blue smoke, dynamite smoke, would just hang there. You would feel your way in. The only time you got air was when you shoveled coal.

Modern mines employ larger fans and baffles (either permanent or temporary) to ensure proper ventilation throughout the mine. In some cases, however, this was still insufficient to prevent asphyxiation from fumes released when electrical cables burned:

At Westover [he] was involved in a fire with an electrical cable. He got a belly full of that, and it killed him.

Just as the shuttle car was comin' to get a load, the car hit a loop in the cable and grounded it. It went off just like a candle at Christmas. We had to fight to get behind the canvas [temporary ventilation baffle]. I couldn't half-breathe for a coupla days.

The level of dust in the mines rose sharply with the advent of full mechanization (Dix 1977). Although most machinery was equipped with spray nozzles, the sprinklers were ineffective before the 1969 Mine Safety and Health Law:

When I was second shift foreman at Sterling we were usin' chain conveyors. [The men] didn't like it because they got so much more dust. [The company] was supposed to water it down and never did. The coal company was supposed to have pumps and hoses put in to run water in. At Westover it was the same way. They were supposed to have sprayers on the roof of the main haulageway to wet the coal down [in the cars] before they took it out, but they never worked.

It was so dusty when the machines were runnin' that you couldn't see from me to you. They had sprays of water on the machine, but still it never held the dust down.

Those Lee Norse operators, they ate the dust! And you couldn't use that respirator—it would just clog up.

The [Col] Mol or the [continuous] miner make a lotta dust, and I was face-bossin' . . . The miners had about ten sprayers, but you'd still have dust. Then we started to load rock with miners —for headings. The dust was so thick you couldn't see men five feet in front of you.

Talk about dust! At Barnes's 20 the dust would be so thick that all you would see of your buddy was his light. If the sprinklers clogged up in those Wilcox machines, they wouldn't bother to clean them. They *had* to get coal out. Production, production, production!

The dust level not only intensified the explosiveness of methane, it also presented a health problem. Black lung was the end result of a worklife spent in dusty conditions. What is generally labeled as black lung is possibly one of two conditions: silicosis or pneumoconiosis. Silicosis arises from laceration of lung tissue by silicon or silicate dust. This is most likely during rock cutting operations such as the excavation of main entryways; however, operations like cutting coal, rock dusting, and using sand as a friction material on the rail lines can also bring it about. Pneumoconiosis is sometimes called "miners' asthma." It involves the collection of coal inside the lung. As one former miner told me: "It only gets worse until it finally kills ya." He had described his condition to me earlier in these words: "My doctor told me, 'Your lungs are like a piece of soap that you took and rolled in the dirt. Pieces of grit and gravel would be stickin' all over it. That's what your lungs look like.' [He] told me that the thing that made my lungs so bad is that they hadn't had a chance to develop before I was exposed to the coal dust. I was scrapin' up bug dust at age thirteen and usin' the cutting machine at age sixteen."

Although there was a miner's hospital in Spangler, this facility was too far from many mines to be of much help in an emergency. Men who worked in such towns as Mentcle, Heilwood, and Nanty Glo could expect their emergency help in the way of an ambulance from Johnstown, which was nearly twenty miles away. Often ambulance drivers were not familiar with mine locations. One man told me of his personal experience with emergency service: "I was hurt at 10:00 and didn't get to Johnstown until 5:00. The ambulance driver said he couldn't find the mine. My urethra was severed, and I had to go to Baltimore to have it reconstructed. [This was in 1939.]"

Another circumstance that exacerbated an already dangerous situation was shift work. As the scale of mining increased, necessitating the continuous running of pumps and fans, operators found it advantageous to have coal mined around the clock. Contracts began to include the stipulation that men work during a different eight-hour period (shift) every two weeks. The effect on alertness was only mentioned by two men, one of whom told me: "You was like a zombie."

The former mineworkers with whom I spoke were acutely aware of the dangers involved in their workplace. Yet, they were not deterred from working there; I wondered why. I also wondered how the women felt about the situation. The answers were neither simple nor obvious. The more I heard and thought about and tried to understand the natives' treatment of danger the more strands I could identify in the complex weave of emotional practices.

8
Emotions Related to Danger

The Reilly Mine Explosion—November 6, 1922

The blast could be heard for miles around,
In the quiet little mining town.

The Spangler streets were filled with wives,
Praying their husbands would be found alive.

Thirty-three prayers were answered that day,
Seventy-nine men were taken away.

November sixth will be remembered for years to come,
By families and friends who lost their loved ones.
                                        —*Beth Farrell*

Verdict in Spangler Disaster

Coroner's jury finds mine management responsible . . . insufficient number of firebosses employed, open lights being used in the presence of dangerous gas and inadequate ventilation.
                        —*Mountaineer Herald, November 23, 1922*

Were Miners Afraid to Go Back?

They weren't afraid. The miners knew they had to go there, and they trusted each other and the coal company.
                                        —*former miner, June 3, 1988*

Word was that the operators would have to be more careful, have better fire bosses and so on. So I decided to go back in . . . I felt O.K.
                            —*Reilly disaster survivor, June 29, 1988*

What is the relationship between the mine community's feeling about danger and the companies' responsibility for safety? I found in the discourse of the mining community very little evidence of moral indignation toward the companies concerning the dangerous conditions in the mines. I came to understand the reasons for this surprising phenomenon as I learned more about how miners and their wives had perceived and dealt with the threat of danger in the workplace, the nature of their social practices, and how these practices were likely reproduced intergenerationally. These attitudes and emotional practices actually enabled companies to scrimp on safety.

Although I found a high degree of variability in reports of worrying among the women and men in mining families, I was surprised at the predominant response—most accepted the conditions of mine danger with little or no worry. When women talked of the emotional practices that arose from their family or community experiences, they most frequently mentioned that it was something "you grew up with," "took for granted," and "were used to." So persistent was this type of response that it might be considered the cultural norm. Through my analysis of their conversations with me I came to understand how the parents' philosophy, reinforcing statements, and calm demeanor taught children to accept the conditions of their father's employment.

Calm behavior of parents with regard to mine danger was mentioned by both men and women. One woman told me: "[My mother] never worried—maybe that's why *I* didn't." Another woman told me: "I never worried a bit about him being hurt. Even when he was late I just assumed he was working a double shift. The same thing was true for my mother. Both my father and then my husband liked their work. They didn't seem afraid so I wasn't afraid." Males were aware of parental demeanor both at home and on the job. A man described his relationship with his father during his two-year apprenticeship: "Sometimes the props would start to splinter—you could hear them ping. My dad would say: 'Don't worry, the roof'll hold.' Then we'd set another prop next to it. My father never worried about it so I never did either. You were scared at first, maybe, but after awhile you didn't worry." This account is very similar to one recorded by Cooper (1979a:13):

Well that day, we had so many supports pulled out that my dad couldn't understand why the room hadn't caved in already. We kept at that job for a week or so. The roof was so bad by then that I could hear the coal cracking, and as the ribs started to buckle, lumps began to shoot across the room. The timbers were splitting and crumbling all around us and the whole place was rumbling like thunder. I kept saying to my dad, who was working very calmly with a pick and shovel, "Dad, don't you think we better get out of here?" Finally, just as I thought I couldn't stand it any longer, he got up, turned to me and announced: "Well, I guess we better leave." Just as we got to the main heading, the whole thing let loose.

This process of learning from parents' demeanor resembles what Bourdieu (1977, 1984) has called reproduction of the *habitus,* i.e. the unconscious learning of a matrix of unconscious perceptions, appreciations, and actions. This might explain why so many informants were unable to articulate the reasons they didn't worry about danger. Their statement of "You just grew up with it" might be better understood as "You grew up with people accepting it."

In addition to learning from demeanor, a conscious philosophy, couched in phrases, was likewise passed down from one generation to the next. There were two major messages of this philosophy. The first was that danger is ubiquitous in the world (so why worry specifically about the occupation of coal mining?). As one former handloading miner put it: "Sure it was dangerous, but so is walkin' across the street. You could get hit with a car . . . People talk about how dangerous mining is, but look at the steel mills, now they have those high voltage cables runnin' everywhere. I'd work in the mines before I'd work in the mill." A widow recalled: "My husband [killed in a rockfall] used to say: 'When you're careful, nothing can happen. You can get killed on the highway if you're not careful.' " Another miner's wife told me: "I told people [who asked me 'Can you sleep when your husband is on the night shift?'] and myself: 'It isn't any worse than goin' out in the street. Accidents can happen anywhere.' " Several women said of their husbands: "For him it was a job, just like any other." Perhaps not so ironically, the childhood environment in which most of these people lived was filled with danger: railroad trains, abandoned mine drifts, coal tipples, burning hills of mine refuse, mine trolleys, and for those who grew up during the great strikes of 1922 or 1927 or the time without union, the Coal and Iron Police or the company police. Perhaps for them, the world *was* perceived as universally dangerous. Many informants related stories to me from their childhood regarding dangerous activities and its lucky or tragic consequences.

The second message of the mine community philosophy was that fate determines when, where, and if you will be hurt or killed. This was stated overtly by several informants: "I don't care where you are, when it is your time to go you'll be where you're supposed to be." And: "I think that a person has a time when he's going to die, regardless. It wouldn't matter if there were seventeen doctors at his bedside. I guess I'm a fatalist." Most of the people I interviewed were devoutly religious and expressed their ideas of fate in terms of the intercession of God or a guardian angel. A miner's wife told me: "It's all in the hands of the Lord—that's what I said when my husband died so young." A former miner said, "I never got hurt in the mines except for some broken fingers. The Lord musta been watchin' me." A niece of my paternal grandparents told me that my grandfather often used a Slovak expression that translated roughly as: "When your name's written up there [in heaven] there's nothing you can do about it."

The more common means of expressing this premise among older mining community members was in the form of stories of fate, stories in which a person either avoided or could have avoided tragic circumstances. For example, one woman told me: "My brother was in World War I. He was supposed to be moved to the front line, but the only thing that saved him was the flu—he couldn't go up. He lived to be ninety-two years old." Another woman asked me: "Did you hear about those coke [oven] fires in Pittsburgh today? I just phoned my daughter, and she said that her husband just got home. They started right after he left, and he worked right there where the fires were." An older Slovak woman giving people a tour of St. Michael's Church in Cambria City (a former steel and coal town near Johnstown) told this fantastic story about a man changing the lights that ringed the inside of the church, high above the pews. She said: "Once the man fell off the catwalk, but he didn't get hurt because his shirt caught on the wing of one of the angels [statues] in the sanctuary. He didn't even have a scratch on him because the angel caught him."

Men's tales generally dealt with their own episodes in the mines. For example, one retired miner told me:

In the late '30s a man from Portage asked me to come to the Sonman Slope mine to be an apprentice brattice-man—a company man who puts up brick walls [in the mine]. I showed up there, and I didn't know anyone. They put me with a man from Benscreek to load coal . . . The next day, though, he didn't show up. I didn't know what to do so I loaded five cars alone. I looked at the weigh board on the way out of the mine. It showed five cars of rock taken to the dump. Someone, loadin' rock, must've exchanged checks with my car. I never went back. Within a few months there was the Sonman Shaft explosion. I read the list of those who died and saw the name of the brattice-man who I would have apprenticed with. If I hadn't quit, I would've been among them.

One feature of this style of story was spotted by a man about forty years old, the son of a former handloader. He noticed that most of the stories of fate involved *fortunate* circumstances. Using his tip, I analyzed the stories I had collected to that point in my fieldwork. I found eighteen stories describing fortunate circumstances and five describing unfortunate circumstances. This suggests to me that a possible use for telling these stories was to reaffirm the premise that fate determined death, and moreover, that fate (or God) was kind.

These could be described almost as a celebration of the existence of fortunate fate, something worth marking in a world of ubiquitous danger. In one case a woman felt the urgent desire to interrupt my conversation with her husband to tell me a story of fate relayed to her by a friend: "[My friend] said that her son was almost on that flight that crashed. She said that he's supposed to go to England, and he decided to wait until after Christmas. If he had gone earlier he mighta been on that flight. That was about the time he woulda been comin' back." Thus, stories of fate, predominantly with favorable outcomes, were told with relish and listened to eagerly. These "normal" emotional practices, then, included using specific discourses to give shape to one another's perceptions of a dangerous, yet positively, fateful reality.

In summary, then, the philosophy in dealing with the aspect of mine danger has two basic premises that were succinctly expressed to me by a wife of a former miner: "I guess the reasoning is: No matter what, accidents happen, and if something's going to happen, it will!"

Here appears to be an example in which culturally shaped practices and philosophy can affect bodily feeling states. The demeanor, the philosophical messages, and the fateful stories act in a fashion to attribute meaning to reality. Without them, children might learn a different, more frightful meaning of reality and consequently have anxiety about the danger of the mines. In stating this, I am reminded of Beck's work on depression (1972). He emphasizes the importance of maintaining the culturally normal process of selection and evaluation of antecedents and consequents in attributing meaning to reality. In his opinion, depression and other "emotional disorders" can arise in people when their meaning of reality is skewed away from the norms of their culture. One informant noticed the connection between the lack of appropriate practices and undesirable feeling states: "Some women were upset and anxious all the time. I guess they thought about [mine danger] all the time."

I detected competing messages about reality that threatened the meaning assigned in mining families. One woman admitted realizing that "It *is* dangerous if you think about it. I think that it's listed as the most dangerous occupation." Despite this realization, she was still able to convince herself that it was no more dangerous than other jobs. However, this competing view may explain the widespread practice of telling fateful stories; it may be a way of reinforcing the philosophy in the face of competition from other messages. These practices used to reduce anxiety and grief were not always effective. In some circumstances, worry *did* occur.

Under exceptional circumstances virtually all miners' wives worried. A sizeable group of women experienced feeling worried on a daily basis.

One of the most commonly mentioned circumstances for worry was the case of the husband working overtime. Policies by which mine officials notified a wife of her husband's extended work period varied. In some cases men were able to send messages home to their wives to let them know that they would be staying for another eight-hour shift, especially if the mine was close to home. In other mines, even communication by phone was absent. Women pointed to the inability to find out when their husbands would be late as the source of the most intense anxiety in the mines. A Mentcle resident told me: "I especially remember the one night when he didn't get home because of snow. I was down on my knees all night that night—I paced and I prayed . . . That night I was practically hysterical. I kept praying: 'Lord, let everybody be all right.' It was about an hour after daybreak that he came home. I was so relieved when he came through the door I wanted to faint. I was sure I was going to get bad news that night." Similar sentiments were voiced by a St. Benedict woman whose husband worked out of town:

When there was an accident you worried or if they didn't come home when they were supposed to—that was a time you walked the

"If they happened to work a little over- time, the first thing you think is that something's hap- pened."

floor. He wasn't able to let me know if he was stayin' for a second shift. I understand now they have telephones to call home, but that didn't go on then. Even when [my husband] got hurt—they thought that he broke his back—I didn't find out. A neighbor who worked [there with my husband] set [my husband's] bucket on my front porch and said: "[Your husband] got hurt and is in the hospital in Johnstown."

A Nanty Glo woman told me: "If they happened to work a little overtime, the first thing you think is that something's happened. I even got my Daddy-in-law outta bed once thinkin' my husband got hurt. He said: 'He's probably wor- kin' overtime and there's no way to let you know.' I didn't have a phone. It turned out that he worked a double shift. It's a constant worry on ya." Another St. Benedict woman once found out that a late husband *did* mean that he was in a dangerous situation: "If he was ever a half hour late I really worried. I even called a boss once and got him out of bed at 2:00 in the morning. That boss said he was workin' late. Something happened just as [my husband] was leavin' his shift, so he had to stay. He was the only one in the mine. Nobody else knew he was the only one in there!" Times of accident or disaster were, as one might expect, occasions for anxiety and grief. A woman who lost her husband in the 1940 Sonman disaster told me of her feelings:

Around noon on Monday a whistle started blowin', and it just kept goin'—it wouldn't stop. We could tell it was coming from the direction of Sonman. A man came by and said that people were runnin' around like crazy up at the mine. So we went out on the tracks. There were people all over the hills by the mine. I was lookin' for my husband. Every once in a while some men would come out of the mine. I said, "Look, there's Walt!," but it wasn't—only someone who looked like him. The priest was there, and he went to talk to the rescue team. He told me that they were goin' to get the men out. Later, the priest told me that he knew they were dead but he was afraid the people would have rioted if they knew. There were about six company officials there—they would have killed them.

Someone told me that my husband had told him that he put his key under the car horn. We looked and found it. I thought maybe he'd come out and decided to go back in and help with the rescue, but after awhile I decided he didn't want that key in the mine with him that morning —he'd lose it.

That night no one slept. I kept lookin' and lookin' out the window. They didn't get them out until Wednesday. They took the bodies to the firehall and tried to clean them up. One family lost two sons and a father. At the funeral home those people were cryin' their eyes out. Everyone in town was affected; those who didn't have family killed, had friends. The whole town mourned. Businesses closed, the schools closed.

I was a wreck for a year after that. I swore that my son would never go in the mines.

Many times I would open up that cedar chest and take stuff out until I saw the ribbons from the flowers. Then I'd start cryin' and cryin'. You never forget—I don't care if I'd live to be a hundred. [Tears rolled down her cheeks.]

Another woman in her eighties recalled an episode from her childhood. I spoke with her twice and she chose to tell me this memory both times: "My dad and two brothers were killed at Derringer's Mines between Garmantown and Moss Creek. I was only seven then [1908], but I can remember the caskets in the house and my mother cryin'! They had her strapped to a chair." And she also recalled the Reilly Mine Disaster:

At the time of the Reilly Shaft explosion we were livin' in Watkins. We were just married. A girl came to Watkins and said that there had been an explosion. "My God!" I said, "That's where my husband's dad and brother work!" I went down to Spangler. My husband worked at Barnes's. All the miners there quit and came out of the mine when they heard the news. My husband just threw his dinner bucket on the porch and run down to Spangler. He wasn't allowed to go in and try to rescue his brother and father. He stayed all day, and I was worried about him. His mother said that she knew something was goin' to happen because there was a deep red sky that morning. I was bakin' bread for her because she was goin' back and forth to the tipple.

A retired miner recalled the Reilly disaster: "I sat down there at Reilly's Mine and watched. There was hysterical cryin' all over. They brought out men burnt black." Another retired miner recalled his mother's dealings with mine death: "My mother's both husbands were killed in the mines. She almost went crazy. She went through awful, awful, awful pain. And she didn't get any compensation, but that's the way it was." Women were generally more worried about their sons working in the mines than about their husbands. One woman stated it this way: "When our boys went in [to work at the mine] it was really more of a worry. We didn't want to see them go in the mines, but you couldn't tell those kids. They saw that the money was good. These kids would stop for a beer on the way home. My husband never did that."

A circumstance that held special weight with the first generation Catholic immigrants, although it has lost some

significance with their offspring, was the case of men working on holy days of obligation. It was impossible to petition
the Lord for a husband's safety when he was flagrantly com-
mitting a sin. It was like challenging fate. A daughter of immigrant parents told me: "The Italian women tended to be
superstitious. You'd hear: 'If he hadn't gone that day, but *he*
had to. . .' It would be a holy day." A retired Catholic miner
told me: "A man got hurt in the mines the Saturday before
Easter and that woman was so worried on Easter Sunday they
both died—yeah! . . . She didn't want him to go to work on account of it bein' a Catholic holiday. He got his arm hurt and
was in the hospital. I don't know if he had a heart attack
or what, but he dropped over dead, and she was dead within
ten minutes.

And about himself he said: "I worked once on Easter Sunday! I really *did* worry then 'cause it wasn't right." He indicated that more than religious belief might be cause for
worry: "Grandma McCormick wouldn't let her grandson go
to work on a holy day of obligation such as the Ascension.
She thought it was more dangerous. Sometimes it *is* more
dangerous when there's not a full crew there. Sure it's more
dangerous 'cause there's nobody around ta give you a hand."

Another circumstance that was conducive to worry was not
saying goodbye to the husband before he left for work. The
thought of his subsequent death led to anxiety. This made
family disputes more problematic. The wife of a retired
miner explained it to me: "[My husband] and I had our fights
just like anyone, but when we stopped fighting it was over. I
think that there was only one time when he left while we
were still mad at each other. And it was terrible! Then you *did*
think about it all day." In fact, I heard several widows express
directly, or tell stories of, moral indignation over a woman's
practice of not kissing her husband goodbye because of a
fight or because of her desire to sleep in. Thus, at least for
some women, there was a moral code involved with the ritual
of departure for the mines, the breaking of which should in-
duce shame in the woman.

While some women worried on the occasions I've just described, other women felt worried on a daily basis. Several factors seem important to these women, and to other mining community members, in explaining why some women worried more than others about their husbands' safety. I've already described parental messages (both verbal and nonverbal) that might reproduce calmness in descendants. Women also differed in their understanding of the dangers involved in mining.

For most women an understanding of coal mining was obtained through four sources: overheard conversations among men, direct conversation with a miner, visiting a mine, and experiencing injury or loss of a family member. All of these sources were more available to daughters of miners than to daughters of non-mining families. It was in the household and neighborhood where the women experienced not only emotional practices regarding the danger but indications of the danger itself. As one woman said: "I think that anybody who has people in the mines is worried. You *think* about it. They come in and tell stories. I would hear my son tellin' his dad—not me. At the time you think about it. I think that my mother worried about men in the mines. There was always talk among the men over there [at her home]. You took it for granted."

Direct conversations with men about the danger of their job was less likely, but possible in some families. Many women, regardless of family background, obtained the chance to go inside a mine when men were not working. Mine engineers or foremen could get permission to bring in women kin and their friends. In the 1960s a tourist mine opened in the area and women also took this opportunity to experience the environment in which their husbands had been working. In every case, it seems, the experience heightened women's concern for their husbands. For example, I asked a woman from Nanty Glo if she ever worried for her husband's safety. She replied: "Oh yeah, I knowed how dangerous it was. Me and another lady went in with her husband to see what it was like.

And my uncle ran a house coal mine outside Philipsburg . . .
It's a constant worry on ya." Another woman told me: "[The
danger] was always on your mind. I guess we were a mining
family and you accepted that. After I was in the mine once I
said [to my husband]: 'We'll save our money and retire as
soon as we can.' I couldn't believe it! My brother was mine
boss and he took us in once when the mines weren't workin'.
We went right in to where they worked. We were scared to
look." A third woman told me: "I never went in the mines. I
did go into [the tourist mine] after my husband died. That
was terrible, the conditions they had to work in." A fourth
woman said: "Ten years or so ago, we went to [the tourist
mine]. That's when I saw the inside of a mine, and I thought
'So *that's* what it's like.' " The same woman, after asking me
what other women said and listening to a story I had heard,
said: "I believe it! She must have known more about the mine
than I did."

The familiarity with mine danger did not always heighten
anxiety, however. One miner's wife discounted the notion
that a mine death in the family would contribute to later
chronic anxiety: "I knew others who had a father or a father
and a brother or two brothers killed in the mines and not be
upset about their husbands being in the mine. I think it depends on the personality of each person."

The statements from or about women who felt anxious
about their husbands often describe feelings of powerlessness
or of anticipated tragedy. Like an observer whose anxiety
level rises as they are forced to watch others walking on the
edge of a dangerous cliff, they were forced to watch their kin
depart for the dangerous mines. Those in the dangerous conditions had some sense of the riskiness of their actions, some
sense of control over their bodies by which to avoid risky situations or to be alert and vigilant. Those who had to wait and
anticipate could only imagine the risk and had no sense of
control over their loved ones' bodies.

Concerning the anticipation, one woman was said to have
prepared daily for the worst by laying out good clothes to put

on at a moment's notice. Another woman was reported to have followed the practice of always getting housework done early in the morning. Each wanted to be prepared to go to the hospital should word come of her husband's injury.

In what ways could women attempt to exert some control in this situation in order to modulate the feelings of powerlessness? Some women with whom I spoke made efforts to exert power through prayer. For example, a former Bakerton resident voiced the following response to my question as to whether wives of miners worried about their husbands:

Oh my God, that was the biggest worry! As a child I used to pray every day that my father wouldn't be killed, especially when he was on night shift. Two boys I went to school with, both 18 years old, died in the mines. They were our neighbors and friends. It was a big worry. Especially years ago when the miners used dynamite, and things were less safe. I went to New York to work when I was 16. The whole time, I prayed all the time for them to come home alive—everyday. Everyone did it.

Similarly, a Mentcle woman told me: "The fears I had were constant. Everyday I prayed." They prayed that God would control situations of possible danger or control their loved ones' bodies. Some women sought control over their sons' or grandsons' bodies—to prevent them from ever entering a mine. The woman whose husband had died in the Sonman disaster stated emphatically that she would never let her son enter the mines. Another woman told me: "Jimmy wanted to go into the mines after he came out of the service. The only other job he could find was bartending, and that didn't pay much. He told me, 'I'm gonna get a job in the mines—look how much they're payin'.' I told him, 'Jimmy, you're not going into the mines! I lost a father and a husband in the mines—I'm not going to lose another one!' " A woman from Emeigh Run told me: "I told my grandson, 'Whatever you do, don't work in a mines.' What did he do?—ended up workin' in that mines in Utah that had that explosion."

I asked another woman if mine accidents caused her to
worry about her husband working in the mines. She replied:
"Oh yeah! He was a motorman, and sometimes he had to
work extra—that would really make me worry. That's why I
didn't want my son to work in the mines." Most women were
hesitant to talk with their husbands about their feelings of
concern. However, some women sought to talk their husbands into leaving the occupation of coal mining:

I remember asking [my husband]: "Why don't you look for a job
above ground—even at five dollars a day?"

[After the week of being buried in roof debris four times] I came
home and asked her, "Do you have enough money saved up for us
to get outta here?" She'd been after me to leave for years.

Goodness knows I didn't want [my husband] to work in the mines,
but what could you do about it? I had a brother who would be able
to get him work in Coreopolis, but when he wouldn't go, what could
you do? I always wanted him to get outta here.

As I heard of these conversations between husbands and
wives, and of unspoken thoughts, I began to see them as
the medium for everyday family politics and as a source of
divisiveness.

When I began interviewing people, I expected the topics of
danger and worry to be spoken about most often. What I
found is that men frequently talked of the dangers lurking in
the mine, of disasters and near-disasters, and of their own
vigilance. They celebrated their own survival within the atmosphere of potential catastrophe. Occasionally they would
speak of their fears or worries, but most men denied ever
having them.

Women seldom introduced the topic of danger or worry. It
was only after several months of interviews that I began to
ask questions directly on those topics. I found that when I introduced the topics, women responded. Most seemed eager
to talk and responded to my questions with an aura of seri-

ousness that reminded me of a diplomatic conference. Often grandchildren were asked to play elsewhere, or we sequestered ourselves in an isolated spot in the house.

Why had they held back talking to me of danger and worry? Why did they take my questions with such seriousness? I have two theories. One possibility is that women may have perceived their concerns as not being newsworthy. In the era and place in which they lived their young lives, women were expected to spend their time in the house, occupying themselves with domestic concerns. In some respects this tradition remains even today. Some women may not have seen their lives as subject matter for history or anthropology. There is another indicator of this: I had some difficulty at first finding women willing to talk about their memories of the 1927 coal strike. A common answer to my request was, "I don't know what I would tell you." Later, I learned to couch my request in terms of asking them questions about their lives. Their seriousness in the interviews about danger might reflect, in part, their appreciation of being taken seriously by someone who wanted to write about them.

An alternate explanation is that discourse between males and females concerning the topics of danger and worry was not an accepted practice in the family or community. In other words, my gender may have denied me access to their spontaneous expression. Later, when I brought up the topic as an anthropologist/historian/author, the context was perceived as more serious and conference-like. Under such conditions discourse was be more acceptable.

I feel that the latter explanation holds more weight, for I noted in people's descriptions of their lives a lack of discourse between miner and wife regarding worry or danger. In general, this appeared to be an unspoken understanding, although at other times it involved a request or order from the husband. What follows are examples of discourse that help to describe this spousal interaction:

I asked one retired miner if he thought the mines were dangerous. He answered: "I wouldn't say they were—that A

[seam] coal that I worked the last two years had bad roof. You couldn't trust it. It was the worst roof I seen. You'd get one roof bolt in and the ceiling would be fallin' down on the other side." His wife interjected the comment: "It's a dangerous job. A lot of men get hurt." The husband then spoke while look-ing at me, but the comment was meant at least partly for her: "Hurt, yeah, but I mean only one guy I remember gettin' killed. The day I started in Viola. a man got hurt and later died."

This reply is interesting on several counts. First, it seemed meant to discredit and stifle the wife's concern about danger. Secondly, in it he has chosen to equate danger with death, rather than with serious injury. Finally, it excludes from con-sideration the death of his partner in a house coal mining op-eration, whom he described earlier as having been killed in an accident with an auger machine.

Another interaction between husband and wife concerned the seriousness of an accident the man had experienced in the mine. Most miners did not play down the seriousness of their mine injuries in their conversations with me; however, I usually interviewed them when alone or in the company of other men. This retired miner, in the presence of his wife, said: "I never was seriously hurt [in the mines]." His wife in-terjected the comment: "You were burned badly that once." He then said, looking at me: "I was scorched a little—an elec-trical fire." She didn't let it lie: "You were out all summer—three months!" She said. I felt uncomfortable about pursuing the subject further. The contention seemed to me to indicate some underlying issue—probably one related to how danger-ous the mines were.

In another instance I spoke with a husband and a wife sep-arately about worry and danger. The former miner told me this when I asked if he discussed the danger of the mines with his wife: "I almost got hit by a rock once that fell all the way from the seam rider [a spurious vein of coal lying above the major vein]. You didn't want to say that at home. It would be negative stuff. I thought that it might bother her more. It's bad enough you remember it yourself—you don't talk about it." His wife later told me: "Danger was a subject they never wanted to talk with you about. If there was a discussion you was out. You didn't need to know . . . It was a matter of I'll stay out of your business if you stay out of mine. He'd say, 'You don't need to know. It's none of your concern.' " Here was a case with the overt understanding that certain topics were not to be discussed between miner and wife.

I asked a retired miner if women used to worry about their husbands being in the mine. He answered: "Oh yes, my wife used to worry about me. She didn't make it a habit of tellin' me or anything."

I asked a widow if her husband had ever worried about being in the mine. In her reply she indicated inter-spousal conflict: "He wasn't the person who would tell ya. He didn't want to upset me to start with—he knew I had always fought with him to get out."

A wife of a retired miner had told me of worrying about her husband for the first time when she received a call telling her he was injured. I asked if she mentioned to her husband that she had been worried. "I didn't really mention it to him. It was just that first week or so that I worried. Then he was off work for quite a while. No, I don't think we ever discussed it. We came to terms with it ourselves [individually]."

I asked the daughter of a deceased coal miner if her mother ever worried about her husband. She told me: "Yes, you could *see* this as he would be leaving for work." Again, any evidence had to be observed in actions or facial expressions rather than in directly spoken language.

The origins of this practice of disallowed discourse seem to lie in male discomfort with women's anxiety or grief. I will present three pieces of evidence for this from informant's statements. The first statement came from a retired miner whose buddy was killed in a roof fall that also left the miner permanently disabled. I had asked him how he felt about the accident: "Real bad! We was—my God!—close. I think about it almost every day. [The accident was in 1941]. The [worst] part of it was his wife had just had a baby. He was going to quit early and bring his wife and baby home. Every time she saw me she started crying so I didn't bother with her much."

Another retired miner told me of reluctance to inform a woman of her husband's injury: "Once at Sterling a man went to move a prop and the whole roof fell on him and slightly injured his buddy. They took the injured man to the hospital, and I got elected to go and tell [the injured man's] pregnant wife. Luckily, his grandfather was there when I got there, and he told her [later]." How this relates to control of emotion discourse is well-articulated by a wife of a retired miner who found herself unable to communicate her worry to her husband: "If I said, 'Please be careful,' he'd say, 'I don't want to hear it.' I realized that I might transfer my fears. I remember one three-week period with no conversation. I knew that he must be under a lot of strain, but there was no way I could help him. I didn't want to transfer my fear. I couldn't share my worry—it was like planting a seed."

Later, I asked the woman why she thought it was that women had remained silent about their worries. She replied: "I can't say why. I know in my own case that I detected early in our marriage that he didn't appreciate my concern—no, I mean he didn't appreciate my voicing my concern about his safety." Her husband corroborated her notion that he would worry if other people feared for his safety: "People told me I was crazy to go back [in the mines to qualify for a pension]. They said, 'You want to get a little more, but you'll get hurt and lose everything.' I didn't want to hear that. That was a horrible strain on me." Another former miner also indicated that he wanted no provocation to think too much about danger: "A week after I left my buddy Mick he got blown up on the longwall [a machine that removed two hundred-foot swaths of coal using a spinning head]. My buddy Jim got his back broke when a rib-hanger [rock on a tunnel wall] came down. Each of these guys was my buddy-buddy. It wasn't a week after I left that each got hurt. I started thinkin',

'What if I was there? Maybe I woulda got hurt, or maybe they wouldn't have.' You don't think too much, it would bother ya."

The point is that it is possible that a wife's emotional expression of worry could have adversely altered a miner's view of danger, increasing his worry as well. Under such conditions, emotional discourse was curtailed by the wife based on the reasoning mentioned by the wife in the above statement or because of a direct request or order by the husband. Other men indicated that they kept worry to themselves either out of concern for their wives' feelings or to avoid conflict. Some men chose to remove themselves from danger, worry, and wifely concern by finding another line of work.

One consequence of this emotional practice is that many miners reserved for conversations with other men their talk about dangerous conditions. This led to a distancing from women in both the physical and emotional sense. Often, men would congregate together in one part of the house or in bars and social clubs, away from women.

Discourse also seemed to be more limited among the community women than I would have imagined. Generally, a wife only talked about her feelings to another female family member: a mother, sister, or sister-in-law. As one woman said: "Other women didn't talk about [danger] to me, either." Many informants, male and female, indicated that females would talk with other wives in the community during times of mine disasters when women met at the mine or at the hospital to determine the fate of their kin. Ladies' auxiliaries of groups such as the church or the UMWA were said to aid a grieving woman, but by providing food and domestic labor rather than consolation, counseling, or emotion talk.

The worry expressed by my informants involved complex cultural and interpersonal features. These worked in two fashions to remove the mine owners from scrutiny. The emotional practices monopolized attention that might otherwise have been directed in a critical fashion against the company, and they were formulated to involve strictly processes internal to the working class.

If one maintained that the rest of the world was as dangerous as the mine, then the company or owner could not be made more responsible for safety. If one believed that fate was kind, then the level of danger was much less important. If the discussion of danger was a source of discomfort or of conflict, then its existence would not be communicated openly. If moral indignation was used by the community to judge interspousal practices such as kissing a mining husband goodbye, then a woman may only have had herself to blame for subsequent anxiety or grief. If everyday politics set a woman's goal at gaining power over her husband's or sons' bodies then she eschewed the goal of collective power to affect the mine owners.

In addition to these practices, perceptions, and attitudes was the tendency of miners to blame themselves for accidents. As I talked to former miners, they gave me the impression of the mines (particularly as they existed prior to 1960), as being both a place of ever-present danger and of constant miner vigilance. Young miners were taught to listen to and watch for what the mine told them. It was as if the mine were populated by living things warning them of danger: cracking mine timbers, chirping roofs, pinging roof bolts, and falling flakes. They gave many examples in their conversations with me:

Sandstone busts, pops, and makes noise; God-darned slate with wet veins in it will fall with no warning.

Pins [roof bolts] give you some warning; they ping when weight's put on 'em. When props [mine timbers] talked to ya, you understood it. They can crack at the top or at the bottom. The floor can heave, too! You'd learn from experience how to read them— nobody could teach ya.

When you was workin' pillars [removing the blocks of coal that were supporting the roof] you'd always hear it poppin' and crackin' someplace.

I worked under sandstone roof at Sterling. There'd be a rip go through the sandstone and everybody'd run. They thought I was crazy for stayin' there. Hell, I knowed the roof wasn't goin' to fall till the posts cracked. All them cracks'd come together. I been sittin' against posts already where they give you a kick.

As I worked, I started to feel some flakes falling from the roof. I left the machine and got out of there. The foreman went after the machine, and the roof fell on him and the machine.

I shut off the machine, and I could hear the [roof] bolts pingin', one right after the other like someone playin' the piano.

You'd know when the roof was gonna fall—the props would snap. I remember the first few times I heard that. I would jump, and my dad would laugh at me.

I heard a 'chip, chip, chip' down the entryways for days. I said, 'This place is gonna cave.' . . . When it came it looked like a big curtain droppin' down. Eight sixty-foot entries all dropped in less than a minute . . .

Sandrock is terribly loud. Most miners would rather work under slate. It gives more warning—[flakes of] shale falls.

A prop wasn't just used for support. The wood will split and crack and give you a warning so you could get out.

I told him it was too dangerous—that I could hear crickets in the roof.

I found the image of mine timbers particularly fascinating — standing there as sentinels to give support and warning. I discovered the image repeated in the dream that a former miner had experienced during a near-death experience [his heart had stopped during aortic catheterization]: "I was on my way out of a mine and there were timbers on either side holding up cross beams as I walked through. Then there were men I used to know—now dead—holding up the cross beams." Thus, the sights and sounds in the mine transformed it into a friendly place where timber, steel, and even rock itself talked to the miners and gave them signs that could save their

"A prop wasn't just used for support. The wood would split and crack and give you a warning so you could get out." Courtesy of the U.S. Department of the Interior.

lives. So the older miners taught the new generation that the mine would look out for them if they only heeded its warnings.

In addition to teaching receptiveness to warnings, the fathers or older men gave younger men adages to help prevent accidents. These taught both specific information about certain conditions and the general message to be ever-vigilant. One former miner put it succinctly: "[Vigilance]—that's what older guys were tryin' to teach you: 'Watch out for the machine; it won't watch out for you,' and 'Don't get in this position; if it happens to stick, there's no place to go!' " I found old, retired miners giving me words of advice, as if we were underground and I were a novice. I'm sure that my ignorance and curiosity to learn reminded them of earlier days when they had taught younger men the lessons of vigilance. Some examples of their lessons to me were:

Never stand between a hoist-drawn car and a rib [tunnel wall, usually composed of the coal seam].

If you take too much squeeze coal [coal in the blocks supporting the roof, compressed by the roof weight] at once, it could fall on you.

Sometimes there was echoes in the roof—it could alert you. Don't tap it too hard. Hold the roof with one hand and tap it lightly to hear the sound.

[When sparks fly], just keep your head down so you don't get your hair singed.

The miner's responsibility for vigilance was suggested by many of the former miners I interviewed. For example, one man told me: "Dad used to get mad at me for fallin' asleep in the mantrip [underground rail cars hauling miners] on the way out of the mine. We were supposed to be watchin' for trouble." Another man said: "Workin' in the mines is dangerous—you have to be on the lookout all the time." And another: "You always have to think before you do anything. I had some close ones, but I was quick." A former mine foreman told me: "Most men [test the roof] as soon as they go in a place. I never had trouble tellin' them." Lewis (1989) has also mentioned the emphasized vigilance of the coal miner from this area.

Vigilance became a necessary practice, not only in the mines, but throughout the dangerous world. The benefit of the mine was that there were so many indications that the mine was looking out for miner's well-being. Not only was fate likely to be on his side, but, if he was alert to warnings, the mine was likewise looking out for him. Although constant vigilance was an exhausting practice, its benefits included increased safety and reduced anxiety.

It is an amazingly short step from a miner's responsibility for vigilance to his responsibility for accidents. This latter responsibility can be seen in the statements of former miners themselves:

Sure it's dangerous, but for a lot of the men who were killed, it was their own fault—carelessness.

Two thirds of the accidents in the mines aren't surprises.

I'll bet 80 percent—*more* than 80 percent—of the accidents are the fault of the men.

It was the philosophy of ubiquitous danger and kind fate along with the recognition of the mine's benevolent warnings to those who were vigilant that made most accidents fall into one of two categories: those for whom no one was to blame (an act of fate), and those for whom the miner was to blame. Thus, practices that were traditionally used by these miners to reduce their anxiety had the additional effect of often absolving the company of blame for mine accidents. There is an important extension of this explanation which I will address in a moment. Before that I would like to suggest two additional explanations for miners' acceptance of responsibility for accidents.

The first involves the fact that many (although certainly the minority) of the miners with whom I spoke came from farming backgrounds. Having worked with older relatives on the farm, they were taught to be careful around draft animals, livestock, and machinery. To be injured there through lack of vigilance was clearly a matter of the individual's responsibility since no owner or company was negligent in maintaining reasonably safe conditions. A possible exception here, though not likely noted by farmers, was the machine manufacturer who may not have provided adequate protection from gears and belts.

The second reason for individual responsibility involves the characteristics of the handloading system of mining. Dix (1977) has described how the tonnage pay system, hard economic times, and the rhetoric both of the operators and of the United States Bureau of Mines interacted to successfully hold the individual miner responsible for his own safety. I will describe each separately.

Hand-loading coal miners were given much control over production decisions in their workplace. They prepared the coal for loading, laid track to their workplace, cleaned up the place for the cutting machine, loaded the coal, and timbered their individual "room" to maintain safety. Dix (1977) suggested that there was always a trade-off for the miner between time spent on production and time spent on ensuring safety, especially when the economy was depressed [for the central Pennsylvania region, that was roughly between 1922 and 1942]. In order to earn subsistence wages when tonnage rates were low and the mines worked short weeks, the miner had to load as much coal as he could. I found, in the statements of many former handloaders, descriptions of dangerous shortcuts taken by them or by fellow miners: shooting from the solid [exploding a coal face that had not been undercut], checking blackpowder and squib shots that did not fire, using short fuses to set off dynamite charges, and timbering improperly.

So in some respects, the system forced handloaders to neglect their individual responsibility for safety. However, was the miner to blame? The companies, in their trade publications, blamed the miners, and, until about 1930, so did the U.S. Bureau of Mines (Dix 1977). Perhaps the miners came to believe that they were almost entirely responsible for their own safety, and perhaps this belief in sole responsibility for vigilance and the propensity for self-blame continued after the Bureau of Mines saw its error in assigning responsibility and after the institution of hourly wage [about 1950 in the central Pennsylvania region]. Such a process of acceptance corresponds to what Gaventa (1980) termed the third dimensional effects of power. This is the process whereby internalization of roles leads to acceptance of the status quo by the dominated.

Miners had been hearing the message of individual responsibility for years. It came from every sector; even John Brophy, the UMWA District 2 president, who had been a miner for many years and who became the socialist challenger of John L. Lewis for head of the CIO, said: "The miner had to develop something like a sixth sense that would tell him when the chances were going against him, and never miss that warning." (1955:118). This doctrine of vigilance was undoubtedly discussed in bars; in fact, this is where I heard several retired miners speak of it.

Even more pertinent to emotional practices is another facet of the phenomena of self-blame. In a situation where miners and their wives were trying to diminish worry, it is a piece of unfortunate logic that blaming the company for unsafe conditions would be tantamount to admitting that there is something to worry about at work. Following this reasoning, the mining community's evaluation of reality would necessarily change. In other words, the mine would then become an especially dangerous place compared to other workplaces, and the members' cognitive messages concerning ubiquitous danger and fate would not work to quell subsequent anxiety. This possibility of challenge to one's evaluation of reality was suggested to me by a retired miner's wife in her description of her husband's return to the mine after a serious accident: "If he was afraid to go back in, then I would have thought that there was something really dangerous there, and then I *would* have worried. Does that make sense?"

Indeed it does.

Note: The poem that begins this chapter is from *Out of the Dark,* an anthology of student writing from Northern Cambria High School in Cambria County. The project was completed under the guidance of Dr. Erma Konitsky in 1975 to commemorate the seventy-nine men who lost their lives in the Spangler Reilly Mine explosion.

9

Dignity, a Complex Subject

The "little man" is the poor man—the guy that had the least
interest in it and the least control over it. You might know the
most about it, but you take the smallest part of the profit.
—*former handloader*

I pondered over this chapter for a year or more, trying to understand the emotions involved in the union reorganization
during the early to mid-1930s. I thought that the notion of
"the little man" would be the key that unlocked the mysteries.
I feel right in that assessment, but I mistakenly thought for a
long while that the overriding issue was power. While power
was an important aspect in the reorganization, it was dignity
that I came to understand as the men's most pressing issue. It
was only after my fieldwork, as I was trying to teach my students about the harm of bigotry, that the issue of dignity became so obvious to me.

Such terms as the "little guy," "the little man," "the working
man," "the miner," seemed to be used interchangeably by the
retired members of the mining community. As the epigraph
of this chapter indicates, these people saw "the little man" as
poor, unappreciated by mine owners, and with little control
over the conditions of production or of his workplace. This
description of the "little man" best describes the coal miners
of the time without union; however, I believe this term was
used to represent miners both before and after this period,
perhaps with different symbolic meaning.

When I began fieldwork I had no designs on determining
the people's key symbol; in fact, I was not certain that such an
approach was even valid as a way of understanding a group.
However, as I reflected on my fieldwork experience I could

see how most of the people with whom I spoke used the "little man" as what Ortner (1973) has called a "summarizing symbol." It represented for the participants "what the system means to them" in "an emotionally powerful and elatively undifferentiated way." It includes "catalysts of emotions." I go further, to say that it includes emotional transferences that hold the group together in what Appaduri (1990) called "a community sentiment."

I will trace the cultural image of "the little man" as it pertained to the dignity of the first-generation European Americans, to the 1927 strike, to the time without union, to the leadership of Franklin Roosevelt and John L. Lewis, and to the recent collapse of the region's economy. I will begin by analyzing the time without union.

In late September 1928, the coal producer's association endorsed a move to operate its mines under the conditions of open shop—they would no longer honor the union as a bargaining agent (Central Pennsylvania Coal Producers' Association 1928). It was the height of the depression in the Pennsylvania coal fields. The reduced demand of a peacetime economy and competition from southern, non-union coal fields teamed to make mining unprofitable. Some mines closed; others, even the captive mines of steel or railroad companies, operated on a limited basis (Seltzer 1985; Singer 1982). With too little demand for coal and too many miners to produce it, the people from the coal company towns worked shortened weeks, sometimes as infrequently as one day in two weeks. At least some of the miners were aware of the glut of men in the coal fields at that time. One man told me: "How they used to fight for cars—there wouldn't be enough cars. I was only a kid, and they would only give me three cars . . . A lot a mines had too many men workin'. They used to say: 'I wouldn't go there—they have so many men that their asses is stickin' out the drift mouth.' "

People told me that the men were subjected to mistreatment by both mine owners and foremen because of the labor surplus and the loss of union representation. The men were threatened with firing, were punished by being given bad spots in which to work, and were often charged with loading dirty coal. One man told me about his brother being fired: "During those times they'd fire you for almost nothing . . . They fired my brother for loadin' coal before the inspector came. Hell, you could wait two or three hours—your day was shot. You had to wait until they said it was OK to load it. They wasn't thinkin' of people who had to live, people who had to start in the morning to get the work done. I had to go down to the boss at No. 9 and get my brother back in. I told him that we needed the money back home."

Among the most numerous stories I heard were of being short-weighed on the loaded coal cars. Some stories were expressed with indignation, others with humor:

Once my dad brought in his own scales and put only lumps of coal in his car and weighed it before he sent it out. He prit near got fired over that. He figured that the company hooked you five hundred or six hundred pounds or more.

With B seam coal it took three cubic feet to make a ton. They short-weighed us for years until the union come back in. In 1929 they got those big steel cars. Then you got cheated five hundred to seven hundred pounds on each car. They were supposed to do car scaling so that you'd know how much the cars weighed empty. The company never lived up to it. After we got a checkweighman we used to talk among ourselves, "Boy-oh-boy, are these cars gettin' bigger every day!"

When we worked without the union we didn't get much wages. There was no checkweighman, and the weigh boss had the side of his shoe worn out from rubbin' the rod to prevent proper weight. I remember loading this *big* car; it had side boards that folded up. And then we chunked it around the edges and filled it up so full that it touched the trolley lines. It must've weighed over three ton. We got credit for less than a ton. From that time on, we never chunked 'em.

After scabbin' you was lucky if you got eighteen hundred [pounds] on a car. After the union come in, they started inspectin' the scales—they were supposed to be down in the mine. There were no scales down there—it was all in pieces. You might as well be pushin' scale weights around on the floor as on that scale arm. Another thing, they never removed the pins from between cars. How can you weigh cars that way? The poor guys loaded the cars for nothin'.

One story was told to me, with variations, by at least four former miners who lived in towns widely spaced from one another: "At Barnes's one man called out the weights of the loaded cars, and another man marked it down. They had blocks under the scales to limit how high the weight would go. He called out "One ton-two." Here a fifteen-ton motor [locomotive] got loose and accidently run on to the scales."

They also told me numerous stories of working for nothing because they were not paid for dead work [cleaning up roof falls, laying track in the room, bailing water, loading impurities] or because they were accused of loading dirty coal [coal with impurities].

I worked hard one day loadin' six cars of rock, and ended up losin' seven cents. Loadin' rock was considered dead work, and you had to pay for your lamp.

We had to do a lot of deadwork. They didn't pay for deadwork once the union was broke. We even went into the hole because we had to buy our own dynamite. For a long time you'd work two or even one day in a two week period. Some of the bosses said: "Ain't you lucky; you have a job."

You could work twelve, fourteen, sixteen hours a day and not make any money. You'd [go in early to] shoot the coal, clean it, and pile it in preparation for loading, so that you could load the most cars in one shift.

There could be serious consequences for not cleaning the coal of impurities: "Then if you got caught loadin' boney, they pushed the car off on a siding and when you came outta the mines they told you that you had to reload the coal and pick out the boney. They'd move an empty car in on a parallel track for you. Or they'd ask if you wanted the car to go over the dump. Then they'd make you take a day off and in those days the mines was workin' only one or two days a week." In other mines the policy appeared even more severe: "The first time you loaded dirty coal you were sent home. The second time you lost thirty working days. During the late '20s and early '30s the mines were only working one or two days a week."

One former miner told me that he thought the miners were only maintaining the mines for the owners until demand for coal increased. This was his logic: "If they operated the mine one day a week they didn't have to pay for maintenance. Rock falls and rib caves would be cleaned up by men before they could mine coal. All the companies did it." In some mines the men worked in conditions of increased danger: "In those days they was short on timber. There was so many men with broken skulls for the simple reason of rock falls from not havin' enough timber."

Besides exploitation in the workplace, there was destitution at home. The tales of hunger and malnutrition included the time without union as well as the 1927 strike:

In the '20s and '30s there would be kids runnin' around with sacks tied around their feet. People would go from one house to another. People didn't have much, but you was always welcome. Even when there was no union, people stuck together. I don't know how people made a livin' then. I remember kids standin' along the street beggin' the miners, "Please mister, do you have somethin' in your [dinner] bucket?"

I remember Hoover time. We gathered up peelings and ate them. We got flour and sauerkraut from the government. We were lucky to have meat once a week. At Milo Park my mother had boarders. A guy lent us from his store—it took us fifteen to twenty years to pay it off.

Families became tied to working for the same mining company because of the company store debt. The man's paycheck went directly toward paying off his debt for food, rent, and supplies. Yet, for many years he did not make enough to prevent himself from going into greater debt. People told me:

We had a long slack time—one, two, three days a week for about seven years. We drawed a lotta damn snakes [a pay statement with a line drawn for net pay]. The company store would take prit near it all. The highest paid got thirty-some dollars, but then you had to pay for your lamp, your powder, the blacksmith, house rent, wash house, doctor, and hospital.

At Barnes's 12 everyone was gettin' pay checks with only one dollar left after the company store bill. This one time a man got a pay check for five dollars. The superintendent said, "Let me see that," and he tore it up.

In the '30s you worked one day a week and then two days the next week, and you only got fifty cents a week. Things was rough in them days, boy.

Power was exerted by agents of the company to control mining family lives in other ways. Company police enforced a curfew and kept an eye out for rebelliousness and for the signs of union organizing.

When I was nine or ten I remember pussyfoots comin' onto the front porch of our company house in Flannigan. They even came into the house. They had whips, and they wanted to know why my dad wasn't at work. They forced him to go to the mine because the operator was afraid that they were organizing.

People wasn't allowed blinds in their houses; the pussyfoots wanted to see who was in the house.

Blacklisting, or blackballing, was practiced when any man was caught trying to organize the union. His name would be circulated to all other operators in the district, and no one would hire him. I spoke with one man who, along with three brothers, was blacklisted for six years. I also heard many statements from throughout the area concerning foremen's coercion of voters. For example: "You was told who to vote for or you didn't have a job. They had guys posted out there, warning you. They'd count people so as to know how you voted."

It was during this time that "the little man" struggled to maintain dignity in the face of destitution, exploitation, and coercion. For most it seemed like a time when displaying any sense of anger or rebelliousness was self-defeating. The younger generation saw this powerlessness as a part of being "the little man"; for example, one informant told me of action by the company police: "I remember one time [the police] got _____ . A brute grabbed him, and the poor guy's legs didn't touch the ground. [The policeman] was beatin' him over the head with a billy club, and blood was streamin' down his face. Nobody was sayin' anything—nobody was doin' anything—he wouldn't fight back. They had so much control over the individual—they had him under their boots. The poor miner was dirty. He was dirty in two ways: he got dirty in the mine, and he got dirty when he got clean, too." Another retired miner told me: "If he was a good straight man and standin' up for his rights and unafraid, he was in trouble." And: "You'd get so bad till you was boiling inside. You knew you were goin' to work the next day and get beat out of something. And there was no one to talk to but another poor man like yourself."

For one man it translated into a hopelessness: "You'd think that you were goin' to grow up to be nothing. There was no light up ahead. Always you had that fear." However, he also reported words of concealed dignity passed from father to son: "[During the time without union] my father said to me, one human being to the other, 'Don't let on that you're a better person than he is—or an equal to him—because that won't work.' " He was being taught that, even though he must submit to the mistreatment, he should not believe that he was inferior. However, most people, particularly those living in company towns like Colver, Mentcle, and Heilwood, did not escape degradation. As a former Bakerton resident told me, "People in the coal patches were degraded." Many were stripped of almost all dignity through exploitation, poverty, and powerlessness. Informants explained to me that these three elements of degradation were interconnected. The people felt that by keeping families on the edge of starvation and in debt the companies could exploit the men without receiving complaints from the workers. Families became desperate to maintain what little they had. This contributed also to the passivity of community members in the face of overt acts of power by the agents of the company.

Yet being passive led to even weaker feelings of self worth. I understand now why one informant, when I asked if he had any words of advice to me about life, said: "Don't stay poor—get out while you can." Poverty was tied to powerlessness and exploitation.

In the minds of many informants the company was involved in deliberate practices to degrade or punish them. For example, many believed that work days during the "slack time" were chosen purposely to frustrate the miner. A Bakerton retired miner told me: "If you worked one day, the company would make it Saturday. It became known as the 'Saturday Coal Company.'" [When I asked if it was just for meanness, the answer was a resounding "yes."] And a former miner from Mentcle said: "If the company was going to give the miners one day of work, they'd wait until Wednesday and have the men work the afternoon shift. The company said that they did it to save power, but they just did it to be mean."

Another miner was certain that the operators colluded to starve the miners: "The operators had an agreement among them: 'You close your place down and starve the sons-of-bitches out, and then I'll close down my place and starve the sons-of-bitches out.'" And several retired miners from Bakerton told of the practice of Barnes and Tucker Coal Company to use deceitful practices to justify getting rid of any miner they wished:

If the company wanted a man out, they'd wait at the drift opening and throw a bucket of boney into the car. The man would be sent home for loading "dirty coal."

"People was afraid. They kep' 'em poor. There wasn't that much work, and they wouldn't give ya the cars."

They gave my uncle the job of finding dirt on the coal. They'd say to him, "Here, John, dump this on so and so's car when it comes out." They'd do it as soon as you were making too much money. This stopped even before the union came in. Men stayed outside and watched for it.

At least two other informants told me that companies deliberately tried to keep people poor. One informant took me by surprise by making a statement, seemingly out of the blue, as we stepped out of his house on to the porch: "Damn it ta hell! People was afraid. They kep 'em poor . . . There wasn't that much work, and they wouldn't give ya the cars." Another man told me: "The company tried to keep men afraid. They found out it was better for them. Men were afraid of losin' their jobs." And according to several informants, the consequence was being owned "body and soul."

During the time of scabs, if it was time to come out of the mines, the company would bring out the coal first. The men came last. Without the union the miners were like slaves.

People were blacklisted. They owned you body and soul. You had to
live on their property. If you didn't buy they put pressure on you.
You couldn't fight back—they'd shoot you—who would help you?

It is in speaking about this time of degradation that older
members of the mining community described their lives with
conclusions similar to those drawn by Lux (1990:198): that
they existed precariously as a commodity that was inconsis-
tent with living a life of full human dignity. What dignity they
could find was within the family and community where they
had some sense of power and order. Their stories supported
the conclusion of Bodnar (1977) that Pennsylvania workers of
times past turned inward to enclave and family because they
felt powerless to affect the world outside. My informants
would likely modify this statement to read "the degrading
world outside." Their conclusions involve what Beck (1972)
has described as a culturally normal process of attributing
meaning to reality by selecting and evaluating antecedents
and consequences. If the consequences of dealing with the
outside world are poverty, exploitation, and powerlessness,
then one realizes that dignity must be sought in sanctuaries
isolated from that outside world. It was here that the image
of "the little man" appears to have been used to pity and to
nurture oneself and others. It is in this way that I understand
the words that father spoke to son: "Don't let on that you are
as good as he is. . . ." It is a way of saying that they recognize
the discrepency in power, comfort, and respect, but that they
need to be mindful of their own dignity within the conditions
of oppression and degradation. It was within the community,
then, that the practices of adults taught children the sources
as well as the limits of dignity—embodied in the image of
"the little man". As Bateson (1972:194-227) has suggested,
metacommunicative patterns of behavior, such as body movement, gesture, stylistic variation in speech, and tone of voice,
could provide messages for children that affect their self-
image and thereby their mental well-being.

According to my informants, many of these community
members did not read English well, if at all. Companies

would threaten them with deportation if they didn't conform to desired behaviors:

A lot of men were afraid of being deported, but the company couldn't. But the miners had to walk the straight and narrow. Those pussyfoots and the coal companies had lawyers, and they had liars.

The way it was—our parents wasn't citizens, and they were afraid of gettin' sent back to Europe. The superintendent, the squires [company justice of the peace], and even the neighbors, the ones we called "Johnny Bulls" [people of English origin], always threatened our parents. They had people scared—like the Ku Klux Klan.

My father wasn't an American citizen, and he was afraid of being sent back to Italy.

They were more easily made to fear their own sense of power and dignity. It could become a liability. For me, this was symbolized best in a story of consternation over gun ownership: "My father was an alien. He had a gun wrapped in a hanky inside the trap door to the attic. I heard that the Coal and Iron Police were searchin' houses, so I took the gun out and hid it. The Coal and Iron Police came on horseback right up on the porch. They came right in: 'We're going to search!' They would barge right in—no search warrant. They'd choose the time and the houses at random. I told my father to get rid of the damned thing!" The gun was hidden and worried about in the same fashion that one's sense of dignity had to be hidden from company agents.

Because they were uneducated, largely illiterate in English, and trained only in their own line of work, these immigrant miners were often afraid to leave the coal fields for the cities to work. As Bodnar (1977) and the Hylan Committee Report (1922) suggested, it was only when they had relatives or friends in the cities that this generation would move during hard times.

Although there was little overt rebellion during the time without union, some power was exerted against the company. Informants told me of some behaviors that fit into the category of "weapons of the weak" (Scott 1985). For example, some men used the situation of company-store debt to their advantage:

Those people weren't afraid [of being fired because of their store debt]—no! no! They had a guy fired two or three times who neglected his job. They said to him: "OK we'll hire you back, and you see that you pay that store bill." He had twelve kids and an eight hundred-dollar store bill.

I told them: "If I load dirty coal, dump it [instead of making me clean it]." My dad had a store bill so I knew that they wouldn't fire me.

They had you by the rear end because you owed 'em money at the company store, but some guys figured out that they could keep their jobs *because* they was in a hole.

Loading "top coal" was another practice used against the company. In many mines, the boney layer appeared six to twelve inches below the top of the coal seam. It separated two grades of coal: the top coal was said to contain more sulfur and to more readily disintegrate into fine grain material. As such, it was much less desirable; if loaded, it could bring down the quality of the product being sent from the mines. Some companies forbade the loading of this top coal, but miners told me of getting away with loading it:

[The coal inspector] saw me and my dad mixin' in top coal with the bottom coal. He never said anything. They had what they called a double operation: you'd shoot the bottom coal down and load it and then shoot the top down and gob it [shovel it to the side]. That took a long time, and you made peanuts.

We wasn't supposed to load [top coal], but we did. One time we had nice top coal—about this thick [ten inches]. We shot down all the coal and cleaned out the boney and threw it in the gob. We threw a little of the top coal on top of the boney to make it look like we gobbed it all. Then we set on the coal we was goin' to load while we waited for cars. The foreman come by and said, "I see you guys gobbed that top coal. You know, I can tell when someone's loadin' top coal just by smellin' it". And here he was settin' right on top of the top coal we was gonna load.

It seems to me that the practices and images associated with being "the little man" changed from the period prior to the time the union was "broken" in October 1928 to the period of living without a union. In the earlier period, dignity extended outside the community. Practices could include confronting operators, foremen, and even Coal and Iron Police.

They still undoubtedly saw themselves as "little men" when compared to the power, affluence, and autonomy of the operators, but they could collectively make themselves heard, could seek to be understood and respected, and could defy the company by striking. "The little man" would act so as to be heard and respected outside the community. Through these practices the younger generation learned what dignity there was in being a "little man." Accounts of childhood memories attest to this:

My father went up to talk to the managers even though they were the enemies. He told them that we would meet the eight dollars a month [rent]. My father said, "This is a lockout," and went to Syracuse and found a job as a meat-packer and sent money home.

Those old-timers, straight from the old country —you have to hand it to them. They went out on strike and stayed out. Not many people today would do it. A lot of what we have today we owe to them.

Then, men were faithful to their union. They were hard times—it was even hard to get a piece of bread. My dad had a lot of guts. He told Peales: "I'm a union man, and I'm on strike. But I need a road to get in and out of town." He faced Peale—the big boy. I guess they admired his spirit and guts. They said yes.

Anyhow, I was a young kid, and we didn't even get a daily paper. You picked up that kind of news by hearin' older men talkin'.

The old guys, I praise them. But they're not around to enjoy the benefits. Some never got a pension or social security.

I seen Caleb talkin' to the Coal and Iron Police, eyeball-to-eyeball. They was puttin' them outta the house, ya know. I was just a boy and that really impressed me.

Unfortunately, the emotional practices among members of the mining community, i.e. treatment of "scabs" and "pussyfoots," began to tear at the fabric of dignity from within the community while the loss of union effectiveness eroded it from the outside. John Brophy, the union leader from District 2, wrote in 1929: "There is a loss of morale that has no parallel in the 38 years of UMWA history. The miners have been defeated in other strikes but never with such devastating results to the spiritual and moral resources of the miners (Singer 1982:280)."

If the image of "the little man" kept community members turned inward with pity and nurturance during the time without union, how did they manage to reorganize the UMWA locals during the 1930s?

Data from several sources indicated that companies went to great lengths to discourage union organization. Former miners and their wives told me of the intimidating company tactics:

Frank Lucas and my boarder who worked at Clymer were both fired for joinin' the union. This was after Roosevelt was elected but before NLRB [National Labor Relations Board] passed. They had to swear in up at Colver.

The miners had meetings in Jew Town until they got organized. The Coal and Iron Police would be waitin' on horses at the Y. My father was an organizer, and they threatened that he was goin' to lose his job.

Records of the police of Bethlehem's Heilwood Division indicated that the companies sent spies to meetings and kept track of those men who were leaders in the movement to organize (files on civil liberties violations). Why did the men persist? The answer may lie in the appeal that leaders such as Franklin Delano Roosevelt and John L. Lewis made to the sentiments of "the little man." These leaders were mentioned by almost all the older generation community members with whom I spoke. Most spoke of them with a certain reverence: "You could see things brighten up when Roosevelt got elected. You could feel it coming—a good reaction. It seemed like all the women liked him. My wife had a big, framed picture of him." "Roosevelt—never be another one like him."

Roosevelt made it a part of his policy to help the "forgotten man" (Lux 1990:148). It was his empathy for the plight of "the little man," his recognition of the scars of wounded dignity, that attracted members of the mining community. For example:

Roosevelt, I think, was doin' good. If it wasn't for him, we wouldn't a got the union back.

The government wouldn't back you up—then Roosevelt gave workers the right to organize. There were parades—everyone was jubilant. The workers held immense meetings. Prior to that there was just spot organizing because the working man didn't have a leg to stand on.

The employer and employee would have each gotten a better deal if they could have communicated their differences even if it meant shouting at each other. They finally learned that. It's what they do today. There's only one man we can credit with the change— Roosevelt.

Dan Moriarity, chairman of the District 2 organizing committee, sent the president the following telegram: "As Lincoln freed the negro of the South, you have delivered the miners out of the wilderness." During the 1977 coal strike, a community member included these sentiments in her letter to the editor of the local weekly paper (Evansky 1978): "God bless President Roosevelt who gave all men the right to live in dignity—all men of all walks of life. And I say that no industry is going to take that away from us."

Lewis was tied very closely to Roosevelt in the minds of my informants. In fact, one person with whom I spoke referred to Lewis as the president of the United States several times before catching himself. This association may be, in part, due to a policy that Lewis's organizers used in the 1930s. They would tell the men that the president wanted them to join the union without specifying which president they meant (Seltzer

1985:137). I asked one retired miner his opinion of why John L. Lewis was so successful. He told me: "They was afraid of his roar—they knew he was tough. And another thing—they knew he could sway the men and could have their places torn apart in a matter of hours. If he had wanted to run for president of the U.S. he would've won." This caused me to wonder if Lewis's thick eyebrows, which give me the impression of a perpetual scowl, were partly responsible for his popularity—making him appear continually confrontive. I asked one informant to describe Lewis for me. His reply: "He had sort of a long face. He was a *big* man—he was almost like—what was his name during World War II?—but his chin didn't seem so long. He wasn't fat either—just solidly *big*." Size and power seemed tied in this informant's mind. Here was a big man to help "the little man." He tied together the concepts of being big and roaring: "[When John L. Lewis came in] was when the companies started gettin' big men for superintendents. That's why they got Bill Lamont [at Sterling]. He could roar! But he never scared anybody." And he saw Lewis fighting for "the little man:" "John L. Lewis fought for the men. He was a rough son-of-a-gun and used some foul language. He didn't care who he was talkin' to."

Roosevelt was referred to as big also. For example, a Bethlehem Steel Co. policeman reported this from an organizing meeting he attended: "James Mark mentioned the fact that all the men working for Barnes had not signed up with the union because they were afraid of Todhunter, the superintendent. He said, 'He is not such a big man; we expect to deal with bigger men before this thing is over. We will be dealing with the president of the United States.' " Lewis (1930) also portrayed himself as someone who understood the mine worker in a way the rest of Americans did not: "I think the American public does not understand the mine worker and never will. It doesn't understand how a coal miner thinks or why he thinks that way." Myerhuber (1987:90) pointed out that Lewis's position of "No Backward Step" during the 1927 coal strike was "probably unrealistic and no doubt contributed to the intransigence of the operators"; however, "by standing firm against seemingly insurmountable odds, Lewis did insure the loyalty of the rank and file."

It seems as if the presence of Roosevelt and a revitalized Lewis were involved in the miner's changing image and discourse of "the little man." which for them became a symbol of rebellion. As one former miner told me: "Men were afraid of losin' their jobs. Then the union came in, and the miner became a different animal—he changed his spots." Another time he told me: "After Roosevelt got in, the working man got to reverse the situation—the working man got to be in charge." Another retired miner told me: "After we got the union, we got pretty mouthy. If they gave us too much hell, we went out on strike." A third man said: "We were all waiting for the organizing. We had our guns half-cocked. That's when the coal companies was cowering in the corners."

It is possible that the self-image of "the little man" allowed for a shift in interpretation with external conditions. In some instances the image may have provided a motivation for action—a search for dignity. There is still dignity in being a "little man" if one does not live in continual fear of losing one's job, if one is not owned "body and soul," if one can afford a comfortable standard of living, if one can vote freely for candidates, if one can negotiate with the company for honest and decent wages.

The hidden sense of dignity expressed by the father who advised his son, "Don't let them know that you are better than them," maintained a spirit that later blossomed with the new atmosphere of labor freedom and leadership provided by Roosevelt. The miner's report that the coal companies were "cowering in the corners" reflected a sense of exhilaration at attaining the sort of dignity and respect that had been denied for so long by the outside world.

In this respect, the process seems to follow the suggestion of Sahlins (1981) that a revaluation of cultural categories occurs as people perceive their worldly circumstances as changing. As one staunch union man told me, "If they had treated people all right, they would never have organized." The mistreatment followed by support from empathetic "big men" led to a redefinition of the symbol of "the little man." It was no longer a term promoting passivity, but a symbol motivating action.

And as Sahlins suggests, this revaluation will likely occur within the logic of their cultural categories. So "the little man" must remain "the little man" even in circumstances that redefine the term. In other words, the mining community's *doxa* will limit the redefinition of their categories because it limits what is thinkable (Bourdieu 1974:166-68). "Little men" may take collective action and regain dignity with regard to the operators, but they will still remain the workers—the proletariat. They do not challenge the capitalist system, for to become the owners would eliminate their perspective of themselves.

Not only did the empowered miners begin to reorganize their union locals, they exerted power in the polling place. Mining towns became ardent supporters of Democratic candidates. Candidates subsequent to Roosevelt were perceived to follow his lead. As several informants told me: "Republicans never were for the working man."

While covering leadership I think that it is important as an aside to consider the unsuccessful attempt by the local union leader, John Brophy, to win support for the presidency of the cio from many miners of District 2. His speeches were often academic, utilizing a historical approach to explain the roots of labor problems (*The Coal Digger* 1928). This logical, but tedious, approach may not have appealed to the "little man" image of the miners despite his empathy for their plight. Also, Brophy was not empathetic toward the miners who had turned inward to family and community: "The Slavic-Americans, as we know them up there (and I wouldn't be surprised if it's true today), were not interested in radical ideas. They were interested in trying to get a little home together, something like that. They were for house and fireside, and mother and that was about it." (1955:215).

This may have hindered Brophy's ability to appeal to "the little man."

The experience of the time without union led the informants to emotionally support the union. The thoughts of the resident who recalled that "You'd think that you were goin' to grow up to be nothing. There was no light up ahead," may seem extreme. But the feelings are supported by statements of other informants explaining their emotional tie to the union. Some said, "The union would fight for you," and "The unions really took care of the miners." Others said:

During the time of scabs, if it was time to come out of the mines, the company would bring out the coal first. The men came last. Without the union the miners were like slaves. When the unions came back in the men came out first—not the coal.

Without a union, a person doesn't have nothin'. You need a union for a decent livin'. They'd have a person work for nothin' if they could do it.

You can't live without a union—they'll do with you what they want . . . When the union come in, they couldn't do with you what they wanted. They couldn't just get rid of you if they wanted.

Without the unions, they controlled your living conditions, they controlled your income—you might as well say they controlled your life.

For them, the union provided hope, advocacy, a decent living, respect from the companies, protection, and some modicum of freedom from the company's control over their lives.

Most of these miners and their wives have remained loyal to the "reorganized" union and to the Democratic party. Many feel that they have been provided with a comfortable standard of living in their retirement. They point to Roosevelt's social security and to the benefits fought for or provided by the union: disability payments, black lung and silicosis payments, and the miner's pension. This comfort and security have allowed older community members to reflect back on the painful times in their lives, and the embarrassing times, and to decide: "Now the story should be told." However, even at this point things are not well. Deindustrialization, deunionization, and the depression of the local economy tear at the fabric of dignity the people have woven.

Although the loss in mine employment has been startling this past decade, a similar phenomenon occurred in the 1950s and 1960s. In fact, the five-county region of this study lost 30,620 mining jobs over the period from 1940 to 1984 (Pennsylvania Department of Environmental Resources 1985). Thus, the erosion of jobs has been a gradual process. Hopes were raised during the coal boom of the 1980s; more recently, however, about seven thousand jobs in the area's coal industry have been eliminated by the closing of fifty-two mines since 1984.

The official state figures for early 1986 showed county unemployment rates of 12.8 for Cambria, 15.8 for Somerset, 13.0 for Indiana, and 16.0 for Clearfield. The Full Employment Action Council, which counts those unemployed individuals who have run out of unemployment benefits as well as those on benefits, reported an unemployment rate in March 1986 of 23.6 percent for the counties of Cambria and Somerset. For that same date they surveyed eight towns in northern Cambria County and found a range of unemployment from 32 to 60 percent, the average being 45 percent. There is currently very little mine employment, and very little new industry has come in to replace it.

During the summer of 1989 an event occurred that symbolized to natives the death of the coal industry. A salvage operation tore up the spur of the CONRAIL line that passed through the Susquehanna Valley from Logan to Bakerton, Barnesboro, Garman, and Cherry Tree. The line had moved millions of tons of coal during the one hundred years of its existence. The removal of rail lines was not new to the area. As a teenager I witnessed the removal of the New York Central spur from St. Benedict. However, the line through Bakerton and Barnesboro was once a vital artery. NORCAM, an organization formed to save the economy and jobs in northern Cambria County, bought the right-of-way to use as a bike trail with hopes that tracks could be re-laid if industry ever returned.

One informant with whom I had spoken many times brought up the topic of industry pull-out several times in a conversation on April 27, 1990. At one point he spontaneously mentioned his reflections on the developments: "Progress comes and progress goes—look at Bakerton, Spangler, Barnesboro." He later told me about his daughter's visit: "I took my daughter from the Cape and her daughters down to Bakerton to see the miners' monument and where the rails were taken up. My daughter said, 'Gosh, Dad, it looks desolate.' 'Yeah', I said, 'The rest of the world abandoned us down there.' " In June I spoke to the author of a book about Bakerton, written in conjunction with the town's centennial festivities. She spoke of sadness tied to the loss of the familiar and to the loss of hope for a revitalized economy:

I almost cried when I went up to the crossing and saw those big magnets. In fact, I had the camera in the car, and I took pictures. But I think I took pictures through tears. I thought this was the saddest day I ever lived, man, to see that history . . . you just felt like, "the tracks are gone; this is it" . . . Everybody was sorta saddened by it . . . In 1892 the tracks were laid through the tunnel . . . We must have hiked those rails [as kids] a hundred thousand times . . . and wave to the engineers . . . As long as the tracks were there, there was a chance of another company coming in. D'you know what it costs to put a rail line down?

Modern times spell the loss of more than rails. Many of my informants are puzzled and distraught by the disappearance of unions and of union support in the country—and in the region:

The young people today are losin' the union that their fathers fought for. They don't appreciate it.

"I almost cried when I went up to the crossing and saw those big magnets. . . . As long as the tracks were there, there was a chance of another company coming in."

I worked in the shirt factory in Hastings in '31, and we fought for a union because we were workin' 50 hours a week for five dollars—that's ten cents an hour. Now they don't have a union there—I couldn't believe it!

I told the young guys up at Bethlehem 33 [mine], "You guys would go scabbin' tomorrow. You have a fancy house, a boat, two cars. . ."

Thus, again, dignity is challenged as deindustrialization and union-busting threatens to eliminate the intergenerational continuation of their values.

In hearing these people talk of their lives and of their concerns with dignity I came to a realization and to a question. I realized that, for the miner, power had to be a part of dignity because the United States' style of capitalism has required that they demand the conditions of dignity. I found myself asking whether workplace dignity could be any more than temporary for working-class people in an economic system guided by the hidden hand of greed. Would a humanistic capitalism (Lux 1990) or socialism have provided a better solution?

10

Reflections upon Leaving

Mining—it could go on and on and on. Something different comes up every time you talk about it.

—*retired miner*

I slammed the front door closed for the last time; the bolt-lock engaged with a loud "clack." Though I would never enter this house again—what had been home for two and a half years—I did not pause for sentimental thoughts. I stepped down from the porch with cat carrier in hand and headed for my heavily laden car. My cat, "Trouble," and I would make the 2000 mile trip to Colorado alone. The stress of Leah's school work and my fieldwork had taken its toll, and economic necessity had provided the final vector toward our separation. She had found a job in Delaware, I in Colorado.

I put Trouble atop the box that lay seat-belted in the driver's seat. I drove off, passing the town's post office, hardware, dry goods store, church, bank, and restaurant as I headed toward the distant U.S. Route 22. Once on the four-lane, I marked my departure from the region when I passed the exit for Nanty Glo. In my mind I heard the words spoken to me many months earlier: "Don't stay poor. Get out while you can." I had seen an opportunity for a full-time, tenure-track job, and I had pursued it. Now I tucked the fieldwork experience back in my mind like a photograph, or perhaps as a home video. The people and places would run over and over on the same reel, giving me the impression that they would always be there, unchanged, should I return. So the ethnographic pastoral still exists to some extent in my mind—what was, had been, and is yet.

My interpersonal bond-breaking had all been carried out over a year-long period of withdrawal from fieldwork. Crapanzano (1977) has pointed out that writing, following fieldwork, serves as a way to reconstitute the anthropologist as academician. Assuming Crapanzano was correct in his assessment, could one remain living among the natives and also write an analysis of fieldwork experience? I had tried this exercise and had found it to be very difficult. Writing had served to reconstitute my sense of academic self. Besides sequestering myself in the house, I had found myself often driving thirty miles to the library of the university campus where Leah was a student. It was during this time that I had greatly reduced my trips to bars, restaurants, and clubs; had begun using interviews for addressing specific issues; and had reduced interactions with neighbors and friends. It had become increasingly more difficult to be an insider during the interviews, and setting up interviews had become more anxiety-provoking. I now realize that in many respects I had left the area long before setting forth for Colorado. This withdrawal procedure included an effort to avoid transferences I had experienced during the previous two years. I had recognized the existence of at least two emotional transferences that had occurred between community members and me. These were unique to the area—I had not felt them among other groups with whom I had lived. I had been especially aware of transferences since my proposal for fieldwork included using them as a technique by which to discover the psychodynamics of the natives. I define transference as an emotional and often cognitive distortion of reality in which an individual interprets a current relationship as if it were an affect-laden relationship from childhood. In one transference there was a sense of anxiety about finding a job. Here I had sensed that people worried for me, made suggestions of where to look for work, and gave me the feeling, even early in my fieldwork, that tragedy would befall me if I didn't quickly find full-time, secure employment. I had become caught up in this and frantically sought jobs beneath my training and ability.

The other transference was a pity by members *for* the unfortunate—a type of unconditional acceptance and an acceptance *by* the unfortunate of comfort from pity. This had been extended to both males and females, but I had mostly noted it for males. It involved those who were out-of-work, alcoholic, nagged by a wife, dominated by a husband, failures in attempts to become more, or abandoned by spouses. I had indulged in this acceptance on several occasions. These experiences have given me the notion, which I have yet to explore in depth, that transferences can be culturally specific emotional practices. More details of my thinking about transferences may be found in earlier writings (Michrina, 1992).

As I drove the long hours across Interstate 70 I thought about the personal questions and issues which I had brought with me to be considered during my fieldwork. Had I learned or resolved anything?

In planning my study, I had proposed that I could gain access to culturally repressed truth through my status as both an enculturated native and as an acculturated outsider. This self-proclaimed position would allow me the best of two perspectives in detecting important cultural data. As an insider, I would supposedly have an understanding of these repressed truths, and as an outsider I would supposedly be able to note their contrast with the perspectives of others.

But was I a native, an insider? For that matter, how could any investigator know this? What were the criteria? I feel that there are no easy answers to these questions. Is childhood enculturation a good standard for determining insider status? Should the *natives* decide my status? Should *I* judge the way I felt? The problem with using twenty years of enculturation as a criterion for insider status is that it fails to account for differences in enculturation within the same society, both inter- and intragenerationally. In my case, my enculturation differed from most other neighborhood boys because of my father's non-mining occupation and because of my parents' long term goals for me. I had not fit in with the majority of my peers; I had learned different values, perspectives and customs. My enculturation also varied intergenerationally.

As I talked with the oldest generation of residents I had soon come to realize that they had been enculturated in a different manner and had experienced vastly different life events than had I. We could communicate, but I had no innate understanding of their worldview. These differences in enculturation call into question my epistemological advantage—the notion that I could understand them by reflection on my own cultural experiences.

Such intergenerational differences in enculturation seem to me to be likely in any culture where rapid change has been occurring. Similar intergenerational enculturation differences seem likely to me for gender and class differences.

If I consider community acceptance, I remember a warmly receptive group. Some had accepted me because they had known me in my youth, some had known my kin. For some, the fact that I had grown up there, had grandfathers who had been miners, and was interested in their past had facilitated acceptance. I feel that they had seen me as a native son or grandson—perhaps a substitute for close kin who had migrated out of the area. I had sensed an atmosphere of trust that often had led to frank discussion.

I recognize *some* problems I had had in interpreting their metacommunicative routines, but undoubtedly less than a complete outsider would have experienced. I feel similarly about my reading of nonverbal communication and transference cues.

I realize that I occasionally had not expressed my own values when they conflicted with those of informants. It is this practice above all else that would lead me to deny my insider status from my viewpoint. I liked these people; I had been warmly accepted by them; I had been tempted to stay among them. But I had held myself back; I had remained detached enough to maintain data-gathering and analysis. I had not participated fully in their lives, and I doubt that I was able to see and feel the world and its issues through a native's skin. However, I had felt emotional transferences from my interactions with them. I had experienced feelings, yearnings that I had not had in my many years of absence from the region.

There is an ironic twist to this claim of insider status. It had helped to change the course of my study. Instead of having the native perspective inside me, aiding the anthropologist in me in observing and divulging the secrets of the area, the internal native perspective began to question the anthropologist in me with regard to invasion of privacy.

During my fieldwork I had been very sensitive to the issue of informant privacy. I had felt guilty for observing casual behavior among natives and for listening to monologues or conversations—all of which I knew would become data in my field notebook. I felt then, and am now convinced, that these people had not realized the nature of my methods. I had not wanted to inform them that their every action and conversation would be recordable data. I had not wanted to hear the words of disbelief. Perhaps it was because my informants trusted me, as they would a family member, to only present what they chose to tell the world, that I had felt guilt. I did not want to betray their trust, and I did not want to use them.

I realize that the issue of deception in anthropology has been addressed in the past, but I worry that the natives' vulnerability to exploitation by anthropologists is still largely ignored.

I am not referring merely to the covert participant-observation that was condemned by fellow researchers in the 1960s and 1970s (Davis 1960; Erikson 1967; Torgensen 1971; Rynkiewish & Spradley 1976). I feel that what makes participant-observation effective as a data-gathering scheme is its reliance on the naïveté of the native. I fear that few if any natives, even residents of Western countries, realize that their every action and every word can qualify as data to be remembered by the investigator and surreptitiously written down later. I also feel that we purposely remain vague on the details of our method in order to insure its success. I have, in fact, been told so by experienced anthropologists.

Is it legitimate to label this behavior "deceptive"? I have wondered what distinguishes it from eavesdropping and spying. It remains cleverly covert despite the fact that we announce to some (or even all) members that we are anthropologists. I have described elsewhere (Michrina 1992) details of what seems to be deception, and I present some issues involved in performing participant-observation.

As I drove through rain in Missouri my mind continued to reflect back on my field work experience.

The question I had kept in mind during the study was whether my interaction with these people was a resource for them and their kin. Perhaps the most frequent, and certainly the most noteworthy, response of informants had been a desire to tell me and the world about the way things had been for them. Sometimes this involved their obtaining a new appreciation for their own lives and history. Often this had seemed to take place in the glow of their feeling appreciated and of being taken seriously as an equal. One informant told me, "Other people might have started this and quit because it was too much work . . . It's sad—this generation knows less about the way things were like in the past. They don't care." Others had felt like authorities or teachers; for example, one informant told me upon my third meeting to speak with him: "You didn't have to bring me an apple."

At other times this appreciation had occurred during the telling. What had appeared to them to be mundane, or perhaps embarrassing, took on new significance: "It used to be you was ashamed to admit that you came from a coal mining family. Now the story should be told—after all, it's a part of our heritage." This process had allowed them to resolve old feelings that lay in the back of their minds like mementos in the bottom of a cedar chest. In the case of a woman whose husband died in the 1940 Sonman Mine Disaster, the process had literally involved facing the painful mementos she had stored: "Many times I would open up that cedar chest and

"Now the story should be told—after all, it's a part of our heritage."

take stuff out until I saw the ribbons from the flowers. Then I'd start cryin' and cryin'." It had been necessary for her to cancel our scheduled meeting until she could bear removing the contents of the chest in order to get newspaper clippings of the event for me. At our meeting she had told me: "My son was only thirteen months old at that time. Now it's forty-eight years later, and I think it's time for him to see [the clippings] . . . I used to pray to God to keep me alive until my son grew up. After he got married I said 'OK, now you can take me.' " Thus, the process had facilitated communication with her son about those times. Two years later, I saw the two of them at the memorial service for the miners killed in that blast.

Not only had intergenerational communication been facilitated, peer communication had also. One retired miner told me: "I was talkin' to an old man in Heilwood, and we were talkin' about where the old spotlight used to be."

Some informants had gained new insights into their past lives. For example, one man was surprised by the number of dangerous situations he had encountered after he had reflected on it. Others had gained insights regarding their spouses, such as a woman, who after hearing her husband's explanation for fearing returning to the mines after having worked elsewhere, exclaimed: "You must be superstitious. I'm hearing this for the first time. You should trust God more."

For many of the informants, talk of the past had aided in their interpretation of the present. This was particularly true of politics. Most of the informants were "Roosevelt Democrats" who explained to me their thinking about the Dukakis and Bush campaigns as couched in terms of the past. I had felt a strong empathy for their political views, a situation they appeared to have greatly appreciated.

Another topic they had interpreted by explanation from the past was the (then current) miner's strike against the Pittston Coal Co. People had told me:

Every other pay we would have a ten percent cut because of competition. The coal companies tryin' to break the unions today should learn from that—it hurts them, too, with competition.

They're killin' the union. I lived to see the same thing before. They say that history repeats itself, and it's true. It happened before, and it's happening now. The mine owners wasn't makin' enough money because of competition, and the miners wasn't makin' enough to live on. If they don't have togetherness they don't have nothin'.

Don't forget to tell 'em that I took a dollar fifty trip to West Virginia. Just like those guys went now, we went to West Virginia [to picket].

In the conclusion of your book you should write about how quite a few of the companies today are involved with union busting. People from this area are going to Pittston, and they said there are company police there. It's starting to sound like it's going to happen again. We're going back to the same thing. It's kinda scary!

An almost universal response of informants was an appreciation of my taking interest in them. This had been expressed verbally and in the many gifts which I had received: garden produce, freshly baked bread, copies of the UMWA Journal, meals, and canned goods.

I had been in the unique position of being able to circulate chapter drafts to several of my informants. I felt that this was an essential step due to the expository nature of much of my writing. If I was going to explain their behavior or feelings, they should be aware of it and be free to comment. There had been a universal eagerness to read what I had written. Several people had begun reading it immediately, practically ignoring my presence. One informant had told me: "I read it before supper. In fact, supper was a little late."

I felt that everyone appreciated having been included in early readings and having been encouraged to make comments. One woman expressed what I feel all others had felt— a thanks for sharing the drafts with them. One former miner who took care to correct what he thought were misinterpretations had told me: "I'm really going to enjoy reading this. I feel like I'm a part of it now." One reviewer, a wife of a retired miner, had told me of promoting my book: "A woman was here to buy carpets, and I was telling her about your book. I'm selling so many books for you! I should be your agent."

My major fear in writing had been that the language style would be too difficult for many of them to understand. For some this was clearly the case. One man had told me that the draft had a lot of big words in it. As an author, I had to choose an audience for whom to write, and since I wanted to be recognized as an anthropologist, I needed to appeal to the academic audience despite my wishes to provide a meaningful product for my informants. To a certain extent my text satisfies the demands of both audiences. The many quotes from informants make up a text embedded within the academic text. A retired miner or his wife could feel that a proper representation of the past was presented by reading those quotes. The unfortunate aspect is that my formal language may hold from their scrutiny my explanation for what has occurred in their lives. In reading the quotes they may get the false impression that I have presented their information at "face-value."

Some of the informant-reviewers had shared the drafts with their offspring. I feel that this process could have only facilitated intergenerational communication:

My daughter from the Cape read those chapters you wrote. She was out here late readin' them. I went to bed, and she stayed up a coupla more hours, so I know she read them carefully. She told me, "This is good, Dad. Boy, you better not lose him as a friend."

I don't think that my grandchildren would believe the way that things were.

[My parents] are always telling me about this guy who comes around. They're happy that someone is taking an interest in their lives and now writing about it.

We have nephews and nieces who have left this area and gotten college educations. You'd be surprised how interested they are in this.

I have a nephew who lives in Texas. He wants to know what coal is. I went and picked him a box of it.

I had also circulated drafts to members of my generation, who, upon reading them, expressed their surprise at what they learned.

Reviewers were stimulated to talk among peers about the book and its issues. In some cases, the drafts were shared. My fears of resentment by the reviewers for the nature of information divulged or for the nature of my interpretation were largely unfounded. One informant found some fault with both my style and method in the chapter on the work ethic; he saw most of this as being due to the nature of "journalism." He noted the change of tone in the subsequent postscript to that chapter and was pleased.

All of my first audience had expressed their good wishes for me and the project. Perhaps these are best summed up by a retired coal miner: "We hope something comes of this for you."

As I drove through Kansas the noises of the car's strained engine, the tires on pavement, and the air whistling past the cartop carrier set up a monotonous drone. My mind wandered to some philosophical concerns about my recent study. The disbelief in the passive nature of these coal miners, which had been expressed by some with whom I spoke, including one son of a handloader, now a staff member with a major university, left a knot in my stomach. Doing a systematic study of people is a pursuit fraught with uncertainties and possible hidden distortions. I had known this from the start; I had tried to consider these obstacles to understanding throughout my field study. Despite my best efforts to remove all doubt, some still persisted.

I reached behind the seat to grab the box of audio tapes. I needed music to drown out these uncomfortable thoughts and feelings. Bob Seger seemed like the best choice. I slipped the tape out of the case and into the player. The beat captivated my body, but my mind continued its reflexive journey.

The major questions concerning the results of my study involved the character of the mine folk of central Pennsylvania and the discrepency between my analysis and reports such as those of Singer (1982), Dix (1977, 1989), and Goodrich (1925), and with the stereotype of the rebellious miner. Most of the men with whom I had spoken had participated in major strikes after the union had been reorganized, some as recently as 1977. How could they be loyal to the company *and* to the union? I had not posed this question to my informants, but I can offer some conjecture on the discrepency.

The analysis is not unprecedented. Kai Erikson (1976) suggested the presence of passivity among the mining families living in communities along Buffalo Creek in West Virginia. John Brophy (1955:315) commented on the passivity of Slavic miners living in District 2 of the UMWA

District 2 families with whom I spoke may represent a unique case—a skewed sample of all miners. The district's older miners are mostly second-generation Americans whose parents came from Eastern or Southern Europe. As such, most are Catholic. The religious affiliation or the mere one-

or two-generation distance from peasantry may affect their emotional practices.

The history these folk experienced also separates them from the men spoken of by Goodrich (1925). This history includes the Great Strikes of 1922 and 1927, the Great Depression, the time without union, and the period of union reorganization. These historical events may have selectively removed the more rebellious miners from the area out of frustration with their powerlessness to prevent company abuses such as evictions, company police, contract repudiations, blacklisting, importation of strikebreakers, and short-weighing of coal cars. One informant told me that his father, who had gotten in trouble with the mine superintendant for supporting the Townsend Act and who was later implicated in an assault on the superintendent, left the region for factory work in New York State because he viewed the Nanty Glo strike, which led to the great 1927 strike, as merely a lockout by the companies who wanted a closed shop. The father never returned. The same man told me: "About one-third of the people got out of this area and out of mining—they were the lucky ones."

Blankenhorn (1924) mentions an out-migration of the most radical miners during the prolonged 1922 strike. I talked with three former handloaders who had left the area in the 1920s and 1930s without returning and to those who had a reputation for being "mean." I did not note a significant difference in the reports of emotional practices among these individuals.

A manner in which the sampling is unavoidably specific is in age. People in their seventies and eighties would have been twenty to thirty years old when the union was reorganized in the 1930s. Thus, people older than thirty-five years during the time of union reorganization are almost entirely excluded; however, I would expect, based on tradition, that the younger men would be the more rebellious.

I found one of my informants to be notably more rebellious than the others—he seemed to consider rebelliousness to be a part of his image. He had been among those who had attended early organizing meetings in the 1930s and later had become the union representative from his company. In his later life he marched in Washington for black lung benefits and to help draw attention to the Pittston strike of 1990.

Loyalty to the company varied among the residents. A significant factor was the nature of the particular company for whom the man worked. Many men had detested Barnes and Tucker Coal Company for its poor managerial practices. On the other hand, most employees of the Peale-owned companies had been very loyal prior to 1927 and following union reorganization because of the company's respect for the miners.

In fact, Rembrandt Peale may have played a major part in setting a tone of respect for miners in this region. As a Wilson Democrat he had urged miners to vote, he had cooperated fully in business unionism, and he had spent much time in his mining towns such as Glen Richey, St. Benedict, Emeigh, and Dixonville. According to the report of one of my informants, Peale had addressed the union local in their hall in St. Benedict in 1927, pleading with the men to return to work. Although others had not been able to remember this event, most had said it sounded in character with the man. Upon his death, a delegation of miners from Nanty Glo attended his funeral in New York, and representatives of the Clymer union local sent a telegram expressing sorrow for the loss of a "fellow worker." Peale's son, Richard, was reported to have carried on the supportive tradition when he took charge of the mining operations.

With respect to the issue of "truth" or "validity," one must keep in mind that I am analyzing people's memories of past practices and past conditions. Needs or experiences of the informants at the time of telling may have influenced what was remembered of the past or how it was interpreted by the informant. Many of my conversations were with people who were looking back from a position of relative comfort and dignity. They were receiving money from their union pension and from social security, frequently bolstered by either black lung or silicosis payments. Many people now owned their houses, had relatively new cars, and had several children living comfortable lives out-of-state. Would this have caused them to look back with a less antagonistic view of the company? One piece of evidence leads me to think otherwise: the informants had been indignant about the company's policies during the time without union. The bad times were still remembered.

I presumed a continuity stretching from the time of the 1927 strike until recent times—a continuity of emotional practices and interpretation of those practices within one generation of mining families. In other words, I assumed that the way they remembered feeling or acting, or the way they felt at the time of telling, is how they have always felt and acted.

There are arguments that question the validity of assuming such continuity. There may have been a "reinforcing and screening apparatus of general culture" (Frisch 1979:76) that has reduced the memory of anger at 1927 coal operators and has reinforced resentment and contempt toward company agents. Are people remembering how they thought about things then? Given the high visibility and everyday contact of company agents with my informants, and the less visible and more protected owners and managers, I feel that the remembered emotional practices reflect those of the time. For many of my informants, such practices were also recalled for the 1977 strike.

Bodnar (1989) also suggests a social construction of memory. According to his theory, individual memories are influenced both by structures of power and by social discourse. In the case of my study it would appear that powerful institutions affected the reality of conditions rather than individual memories of them. Social discourse would not only affect memories but would also affect emotional practices occurring at various times in the past. I feel that these cultural phenomena are related to and reinforce each other.

Grele (1981:43) suggests that the very process of oral history is alienating—causing people to personalize the past and causing them to see institutions and social forces as secondary to human will. Again, the residents of the central Pennsylvania coal fields did not ignore social forces and institutions in their explanations of the past. Their lack of emphasis on the responsibility of coal operators appeared to be the case both at the time of the events as well as in their memories.

I kept in touch with several former informants. One of these was a former mineworker in his eighties who moved out of state to live with his daughter. I phoned him several times in his new home. I told him of the chapter I had recently written about miners' concerns with dignity. He told me: "It seems as though, if you get ten miles away from a coal-mining town, the dignity for the coal miner is lost. They don't know what [the miners] went through or anything." May this book change all that.

References

Agar, M. 1982. Hermeneutics in Anthropology. *Ethos* 8(3):253-272.

Abu-Lughod, L., and C. Lutz. 1990. *Emotion, Discourse and the Politics of Everyday Life*. Cambridge: Cambridge Univ. Press.

Appadurai, A. 1990. Topographies of the Self: Praise and Emotion in Hindu India. In *Emotion, Discourse and the Politics of Everyday Life*, ed. L. Abu-Lughod and C. Lutz. Cambridge: Cambridge Univ. Press.

Arble, M. 1976. *The Long Tunnel: A Coal Miner's Journal*. New York: Atheneum.

Arieti, S. 1972. *The Will to be Human*. New York: Quadrangle.

Barbash, J. 1983. Which Work Ethic? In *The Work Ethic—a Critical Analysis*, ed. Jack Barbash, Robert J. Lampman, Sar A. Leviton, and Gus Tyler. Bloomington, Ill.: Pantagraph.

Bateson, G. 1972. *Steps to an Ecology of Mind*. New York: Ballantine.

Beck, A.T. 1972. *Depression: Causes and Treatment*. Philadelphia: Univ. of Pennsylvania Press.

Black Elk, W., and W.S. Lyon. 1991. *Black Elk, the Sacred Ways of a Lakota*. New York: Harper Collins.

Blankenhorn, H. 1924. *The Strike for Union*. New York: Wilson.

Bodnar, J. 1989. Power and Memory in Oral History: Workers and Managers at Studebaker. *Journal of American History* 75(4): 1201-1222.

Bodnar, J. 1985. *The Transplanted: A History of Immigrants in Urban America*. Bloomington: Indiana Univ. Press.

Bodnar, J. 1982. *Workers' World: Kinship, Community, and Protest in an Industrializing Society 1900–1940*. Baltimore: Johns Hopkins Univ. Press.

Bourdieu, P. 1977. *The Outline of a Theory of Practice*. Cambridge: Cambridge Univ. Press.

Bourdieu, P. 1984. *Distinction: A Social Critique of the Judgement of Taste*. Boston: Harvard Univ. Press.

Brophy, J. 1955. *Reminiscences of John Brophy*. New York: Columbia University Collection (microfiche).

Central Pennsylvania and Pittsburgh Owners Post Reduced Wage Scale. 1922. *Coal Age* 21(4):548.

Clifford, J., and G.E. Marcus, eds.. 1986. *Writing Culture.* Berkeley: Univ. of California Press.

Cooper, E. 1979a. Commodore, Part II: Community Spirit Lives on Despite Closing of Mines. Indiana *Evening Gazette.* Feb. 10.

———. 1979b. Old Time Mining Part II: Many Sons Followed Fathers into the Mines. Indiana *Evening Gazette.* Apr. 14.

Crapanzano, V. 1977. On the Writing of Ethnography. *Dialectical Anthropology* 2:69-73.

Davis, F. 1960. Comment on "Initial Interaction of Newcomers in Alcoholic Anonymous." *Social Problems* 8:364-65.

Dennis, N., F. Henriques, and C. Slaughter. 1969. *Coal Is Our Life.* London: Tavistock.

Dix, K. 1989. *What's a Coal Miner To Do?* Pittsburgh: Univ. of Pittsburgh Press.

———. 1977. *Work Relations in the Coal Industry: The Handloading Era, 1880–1930.* West Virginia University Bulletin Series 78, nos. 7-2.

Dumont, J.P. [1978] 1992. *The Headman and I.* Prospect Heights, Ill.: Waveland.

Dwyer, K. 1982. *Moroccan Dialogues: Anthropology in Question.* Baltimore: Johns Hopkins Univ. Press.

Erikson, K. 1967. A Comment on Disguised Observation in Sociology. *Social Problems* 14:366-73.

Erikson, K. 1976. *Everything in its Path.* New York: Simon and Schuster.

Evansky, M. 1978. The Miners Are Proud. Barnesboro *Star.* Letter to the Editor. 74(11) Mar. 15.

Foucault, M. 1977. *Discipline and Punish.* New York: Pantheon.

Frisch, M. 1979. Oral History and Hard Times: A Review Essay. *Oral History Review* 70-79.

———. 1989. What Do You Do with Oral History: Moving Beyond the Supply Side. Conference. "Discovering Our Past: Oral History and Industrial Heritage." Indiana University of Pennsylvania, July 12 and 13.

Fulton, J. 1890. *Geological Notes on Cambria County.*

Gable, J. 1926. *History of Northwestern Pennsylvania.* New York: Lewis Historical.

Gaventa, J. 1980. *Power and Powerlessness: Quiescence and Rebellion in an Appalachian Valley.* Urbana: Univ. of Illinois Press.

Gearing, F.O. [1970] 1988. *The Face of the Fox*. Salem, Wis.: Sheffield.

Geological Survey of Pennsylvania. 1888. *Productive Bituminous Coal Measures in Cambria County.* Cambria County Mineral Assessment Office.

Gerber, E.R. 1985. Rage and Obligation: Samoan Emotion in Conflict. In *Person, Self and Experience: Exploring Pacific Ethnopsychologies*, ed. Geoffrey White and John Kirkpatrick. Berkeley: Univ. of California Press.

Goffman, I. 1959. *The Presentation of Self in Everyday Life*. New York: Doubleday Anchor.

Goodrich, C. 1925. *The Miner's Freedom: A Study of the Working Life in a Changing Industry.* Boston: Marshall Jones.

Grele, R. 1981. Can Anyone Over Thirty Be Trusted: A Friendly Critique of Oral History. *Oral History Review* 36-44.

Gutman, H. 1976. *Work, Culture, and Society in Industrializing America*. New York: Knopf.

Halberstadt Map. 1901. *Lower Barren and Lower Productive Coal Measures and Pottsville Conglomerate*. Archives: Cornell University.

Hochschild, A.R. 1983. *The Managed Heart, Commercialization of Human Feeling*. Berkeley: Univ. of California Press.

Holdsworth, D. 1989. Cross-cultural Perspective on Coal Mining Settlements. Keynote address for "Coal Patch: A Workshop on Historic Coal-Mining Towns." Johnstown, Penn. June 22-24.

Hylan Committee. 1922. *Statement of Facts and Summary of Committee to Investigate the Labor Conditions at the Berwind-White Company's Coal Mines in Somerset and Other Counties, Pennsylvania*. Archives: University of Pittsburgh.

Jackson, M. 1989. *Path Toward a Clearing*. Bloomington: Indiana Univ. Press.

Jenkins, P. 1986. The Ku Klux Klan in Pennsylvania, 1920–1940. *The Western Pennsylvania Historical Magazine* 6(2):121-33.

Kaufman, G. 1985. *Shame, the Power of Caring*. Cambridge Mass.: Schenkman.

Kerr, C., J.T. Dunlop, F.H. Harbison, and C.A. Myers. 1960. *Industrialism and Industrial Man*. New York: Oxford Univ. Press.

Levy, R.I. 1984. Emotion, Knowing and Culture. In *Culture Theory: Essays on Mind, Self, and Emotion*, ed. R.A. Shweder and R.A. LeVine. Cambridge: Cambridge Univ. Press.

Lewis, J.L. 1930. In Barnesboro *Star* 74(13), Mar. 29, 1978.

Light, L., and N. Kleiber. 1981. Interactive Research in a Feminist Setting: The Vancouver Women's Health Collective. In *Anthro-*

pologists at Home in North America, ed. D.A. Messerschmidt. Cambridge: Cambridge Univ. Press.

Lutz, C.A. 1988. *Unnatural Emotions, Everyday Sentiments on a Micronesian Atoll & Their Challenge to Western Theory.* Chicago: Univ. of Chicago Press.

Lux, K. 1990. *Adam Smith's Mistake: How a Moral Philosopher Invented Economics and Ended Morality.* Boston: Shambala.

Marcus, I., E. Cooper, and B. O'Leary. 1989. The Coal Strike of 1919 in Indiana County. *Pennsylvania History* 56:177-95.

Mark, J. 1941. Correspondence, Archives: Indiana University of Pennsyslvania.

Meyerhuber, C.I. 1987. *Less Than Forever: The Rise and Decline of Union Solidarity in Western Pennsylvania, 1914–1948.* Selinsgrove, Penn.: Susquehanna Univ. Press.

Michrina, B.P. 1992. *Lives of Dignity, Acts of Emotion,* Dissertation, S.U.N.Y. at Binghamton.

Middleton, D.R. 1989. Emotional Style: The Cultural Ordering of Emotions. *Ethos* 17(2):187-201.

Oring, E. 1987. Generating Lives: The Construction of an Autobiography. *Journal of Folklore Research* 24(3):241-62.

Ortner, S.B. 1973. On Key Symbols. *American Anthropologist* 75:1338.

Peale, S.R. 1861. Letters, Archives: Fulton County Historical Society, Lock Haven, Penn.

Portelli, A. 1981. "The Time of My Life:" Functions of Time in Oral History. *International Journal of Oral History* 2(3): 162-80.

Rabinow, P. 1986. Representations are Social Facts: Modernity and Post-Modernity in Anthropology. In *Writing Culture,* ed. J. Clifford and G.E. Marcus. Berkeley: Univ. of California Press.

Riesenman, J. 1943. *History of Northwestern Pennsylvania.* New York: Lewis Historical.

Rynkiewish, M.A., and J. Spradley, eds. 1976. *Ethics and Anthropology.* New York: John Wiley and Sons.

Sahlins, M. 1981. *Historical Metaphors and Mythical Realities.* Ann Arbor: Univ. of Michigan Press.

Sarbin, T.R. 1986. Emotion and Act: Roles and Rhetoric. In *The Social Construction of Emotions,* ed. R. Harré. Oxford: Basil Blackwell.

Scott, J.C. 1985. *Weapons of the Weak: Everyday Forms of Peasant Resistance.* New Haven: Yale Univ. Press.

Seltzer, C. 1985. *Fire in the Hole: Miners and Managers in the American Coal Industry.* Lexington: Univ. Press of Kentucky.

Singer, A. 1982. *"Which Side Are You On?": Ideological Conflict in the United Mineworkers of America 1919–1928.* Ph.D. diss., Rutgers University.

Smith, D.E. 1987. Women's Perspective as a Radical Critique of Sociology. In *Feminism and Methodology,* ed. S. Harding. Bloomington: Indiana Univ. Press.

Smith, H. 1986. *Export: A Patch of Tapestry Out of Coal Country America.* Greesburg, Penn.: McDonald/Seward.

Torgensen, J. 1971. On Ethics and Anthropology. *Current Anthropology* 12(3):327-34.

White, G., and L. Lindstrom, eds. 1989. *The Pacific Theatre, Island Representations of World War II.* Honolulu: Univ. of Hawaii Press.

Williams, B. 1989. The Coal Miner's Culture: Does It Exist? Paper presented at "Coal Patch, A Workshop on Historic Coal-Company Towns," Johnstown, Penn. June 22-24.

Yankelovitch, D. 1974. The Meaning of Work. In *The Worker and the Job,* ed. Jerome M. Rosow. Englewood Cliffs, N.J.: Prentice-Hall.

Yuhas, T. 1975. *Out of the Dark.* Indiana, Penn.: Hallding.

Yuhas, T. 1977. *Out of the Dark 2: Mining Folk.* Indiana, Penn.: Hallding.

Index